Ion Transport in HfO_2
Ionentransport in HfO_2

Von der Fakultät für Mathematik, Informatik und
Naturwissenschaften der RWTH Aachen University zur Erlangung
des akademischen Grades eines Doktors der Naturwissenschaften
genehmigte Dissertation

vorgelegt von

Michael Patrick Müller, M.Sc.

aus Aachen, Deutschland

Berichter: Prof. Dr. Roger A. De Souza
Prof. Dr. Richard Dronskowski

Tag der mündlichen Prüfung: 08.12.2020

Michael Patrick Müller
Ion Transport in HfO_2
Ionentransport in HfO_2

ISBN: 978-3-95886-393-4
1. Auflage 2020

Bibliografische Information der Deutschen Bibliothek
Die Deutsche Bibliothek verzeichnet diese Publikation in der Deutschen Nationalbibliografie; detaillierte bibliografische Daten sind im Internet über www.dnb.ddb.de abrufbar.

Das Werk einschließlich seiner Teile ist urheberrechtlich geschützt. Jede Verwendung ist ohne die Zustimmung des Herausgebers außerhalb der engen Grenzen des Urhebergesetzes unzulässig und strafbar. Das gilt insbesondere für Vervielfältigungen, Übersetzungen, Mikroverfilmungen und die Einspeicherung und Verarbeitung in elektronischen Systemen.

Vertrieb:

© Verlagshaus Mainz GmbH Aachen
Süsterfeldstr. 83, 52072 Aachen
Tel. 0241 / 87 34 34 00
www.Verlag-Mainz.de

Herstellung:

Druckerei Mainz GmbH Aachen
Süsterfeldstraße 83
52072 Aachen
www.DruckereiMainz.de

Satz: nach Druckvorlage des Autors
Umschlaggestaltung: Druckerei Mainz

printed in Germany

D82 (Diss. RWTH Aachen University, 2020)

Michael Patrick Müller

erklärt hiermit, dass diese Dissertation und die darin dargelegten Inhalte die eigenen sind und selbstständig, als Ergebnis der eigenen originären Forschung, generiert wurden. Hiermit erkläre ich an Eides statt

1. Diese Arbeit wurde vollständig oder größtenteils in der Phase als Doktorand dieser Fakultät und Universität angefertigt;
2. Sofern irgendein Bestandteil dieser Dissertation zuvor für einen akademischen Abschluss oder eine andere Qualifikation an dieser oder einer anderen Institution verwendet wurde, wurde dies klar angezeigt;
3. Wenn immer andere eigene Veröffentlichungen oder Veröffentlichungen Dritter herangezogen wurden, wurden diese klar benannt;
4. Wenn aus anderen eigene Veröffentlichungen oder Veröffentlichungen Dritter zitiert wurde, wurde stets die Quelle hierfür angegeben. Diese Dissertation ist vollständig meine eigene Arbeit, mit der Ausnahme solcher Zitate;
5. Alle wesentlichen Quellen von Unterstützung wurden benannt;
6. Wenn immer ein Teil dieser Dissertation auf der Zusammenarbeit mit anderen basiert, wurde von mir klar gekennzeichnet, was von anderen und was von mir selbst erarbeitet wurde;
7. Teile dieser Arbeit wurden zuvor veröffentlicht und zwar in:

   - M. P. Mueller und R. A. De Souza, "SIMS measurements of oxygen diffusion in HfO_2" *App. Phys. Lett.* **2018**, 112, 51908.
   - M. P. Mueller, K. Pingen, A. Hardtdegen, S. Aussen, A. Kindsmueller, S. Hoffmann-Eifert und R. A. De Souza, "Cation Diffusion in Polycrystalline Thin Films of Monoclinic HfO_2 deposited by Atomic Layer Deposition" *APL Mater.* **2020**, 8, 081104.

Aachen, August 2020
Michael Patrick Müller

Parts of this thesis have previously been published or are being prepared for publication:

- M. P. Mueller and R. A. De Souza, "SIMS measurements of oxygen diffusion in HfO$_2$," *App. Phys. Lett.* **2018**, 112, 51908.

- M. P. Mueller, K. Pingen, A. Hardtdegen, S. Aussen, A. Kindsmueller, S. Hoffmann-Eifert, and R. A. De Souza, "Cation Diffusion in Polycrystalline Thin Films of Monoclinic HfO$_2$ deposited by Atomic Layer Deposition" *APL Mater.* **2020**, 8, 081104.

- M. P. Mueller, F. Gunkel, S. Hoffmann-Eifert, and R. A. De Souza, *in preparation*.

The following collaborative work has been published and is not part of this thesis:

- T. Heisig, C. Baeumer, U. N. Gries, M. P. Mueller, C. La Torre, M. Luebben, N. Raab, H. Du, D. N. Mueller, C.-L. Jia, J. Mayer, R. Waser, I. Valov, R. A. De Souza, and R. Dittmann, "Oxygen Exchange Processes between Oxide Memristive Devices and Water Molecules" *Adv. Mater.* **2018**, 30, 1800957.

- N. Kumar, E. A. Patterson, T. Froemling, E. P. Gorzowski, P. Eschbach, I. Love, M. P. Mueller. R. A. De Souza, J. Tucker, S. R. Reese, and D. P. Cann, "Defect mechanisms in BaTiO$_3$-BiMO$_3$ ceramics" *J. Am. Ceram. Soc.* **2017**, 101, 2376.

- C. Kura, Y. Kunisada, E. Tsuji, C. Zhu, H. Habazaki, S. Nagata, M. P. Mueller, R. A. De Souza, and Y. Aoki, "Hydrogen separation by nanocrystalline titanium nitride membranes with high hydride ion conductivity" *Nat. Energy* **2017**, 2, 786.

- M. Schie, M. P. Mueller, M. Salinga, R. Waser, and R. A. De Souza, "Ion migration in crystalline and amorphous HfO$_x$" *J. Chem. Phys.* **2017**, 146, 094508.

The work presented in this thesis was conducted at the Institute of Physical Chemistry, RWTH Aachen University, under the supervision of Prof. Dr. Roger A. De Souza.

Funding from the German Science Foundation (DFG) within the framework of the collaborative research centre "Nanoswitches" (SFB 917) is gratefully acknowledged, as well as computing resources granted by High Performance Computing (HPC) RWTH Aachen University under project rwth0310.

Parts of this thesis are based on data presented in the bachelor theses of students I supervised during this time. I would like to thank

- Katrin Pingen "Cation Diffusion in HfO_2" (bachelor thesis, 2017)
- Alexander Bonkowski "Rechnerische Simulation von Kationendiffusion in HfO_2" (bachelor thesis, 2018)

for their contribution to this work.

Abstract

In this thesis the diffusion of anions and cations in HfO_2 was investigated in detail. Furthermore, the conductivity of HfO_2 was probed and approaches to interpret the results were presented. The diffusion of oxygen in dense ceramics of monoclinic HfO_2 (m-HfO_2) was studied by means of ($^{18}O/^{16}O$) isotope exchange annealing and Secondary Ion Mass Spectrometry (SIMS). All measured isotope profiles showed complicated behaviour in exhibiting two features: the first feature, closer to the surface, was attributed mainly to slow oxygen diffusion in an impurity silicate phase; the second feature, deeper in the sample, was attributed to oxygen diffusion in bulk m-HfO_2. The activation enthalpy of oxygen tracer diffusion in bulk m-HfO_2 was found to be $\Delta H_{D^*} \approx 0.5\,eV$. The diffusion of cations in m-HfO_2 was studied with samples prepared by cooperation partners, utilising a low-temperature preparation method, atomic layer deposition, in order to produce non-equilibrium samples. These were then used in diffusion annealing experiments and investigated with SIMS. The measured isotope profiles displayed two features, attributed to bulk diffusion and grain-boundary diffusion. A numerical analysis produced a bulk diffusion activation enthalpy of $\Delta H_b \approx 2.1\,eV$ and a grain-boundary diffusion activation enthalpy of $\Delta H_{gb} \approx 2.1\,eV$. These values are small compared to other AO_2 systems and the difference is attributed to the structural perturbations in the monoclinic system. A computational investigation of cation diffusion in m-HfO_2 using Density-Functional-Theory (DFT) yielded migration enthalpies for individual cation jumps. Two jumps were found with values comparable to the experiments ($\approx 2\,eV$), allowing long-range diffusion through the bulk. Molecular dynamics simulations in c-HfO_2 with an applied field were able to reproduce the activation enthalpy of bulk diffusion determined experimentally and with DFT. However, molecular static simulations instead produce results much closer to those of other AO_2 systems. A cooperative migration mechanism of oxygen and hafnium vacancies is proposed. The conductivity of m-HfO_2 was studied in dependence of the oxygen partial pressure by means of high temperature equilibrium conductance measurements. In reducing conditions the total conductivity was found to increase with oxygen partial pressure. Numerical defect-chemical calculations showed that singly positively charged oxygen vacancies are likely responsible for this behaviour. In the intermediate oxygen partial pressure regime ionic conductivity dominated. In oxidising conditions the total conductivity increased with oxygen partial pressure due to electron holes.

Kurzfassung

In dieser Arbeit wurde die Diffusion von Anionen und Kationen in HfO_2 im Detail untersucht. Weiterhin wurde die Leitfähigkeit von HfO_2 untersucht und Ansätze zur Interpretation der Ergebnisse vorgestellt. Die Diffusion von Sauerstoff in dichten Keramiken aus monoklinem HfO_2 (m-HfO_2) wurde mittels ($^{18}O/^{16}O$) Isotopenaustauschexperimenten und Sekundärionenmassenspektrometrie (SIMS) untersucht. Alle gemessenen Isotopenprofile zeigten ein kompliziertes Verhalten, da sie zwei Features aufwiesen: Das erste Feature, näher an der Oberfläche, wurde hauptsächlich der langsamen Sauerstoffdiffusion in siliziumhaltigen Phasen zugeschrieben; das zweite Feature, tiefer in der Probe, wurde der Sauerstoffdiffusion in m-HfO_2 zugeschrieben. Die Aktivierungsenthalpie der Sauerstoff-Tracerdiffusion in m-HfO_2 wurde mit $\Delta H_{D^*} \approx 0.5\,eV$ bestimmt. Die Diffusion von Kationen in m-HfO_2 wurde mit von Kooperationspartnern präparierten Proben untersucht, wobei eine Niedrigtemperatur-Präparationsmethode, die Atomlagenabscheidung, verwendet wurde, um Nicht-Gleichgewichtsproben herzustellen. Diese wurden dann in Diffusionsexperimenten verwendet und mit SIMS untersucht. Die gemessenen Isotopenprofile zeigen zwei Merkmale, die der Bulkdiffusion und der Korngrenzdiffusion zugeschrieben wurden. Eine numerische Analyse ergab eine Aktivierungsenthalpie der Bulkdiffusion von $\Delta H_b \approx 2.1\,eV$ und eine Aktivierungsenthalpie der Korngrenzdiffusion von $\Delta H_{kg} \approx 2.1\,eV$. Diese Werte sind im Vergleich zu anderen AO_2 Systemen gering, und der Unterschied wird auf die strukturellen Störungen im monoklinen System zurückgeführt. Eine rechnerische Untersuchung der Kationendiffusion in m-HfO_2 unter Verwendung der Dichtefunktionaltheorie (DFT) ergab Migrationsenthalpien für einzelne Kationensprünge. Es wurden zwei Sprünge mit Werten gefunden, die mit den Experimenten vergleichbar waren ($\approx 2\,eV$) und eine Diffusion durch den gesamten Bulk ermöglichten. Molekulardynamische Simulationen in c-HfO_2 mit einem angelegten Feld waren in der Lage, die experimentell und mit DFT ermittelte Aktivierungsenthalpie der Bulkdiffusion zu reproduzieren. Molekularstatische Simulationen lieferten jedoch stattdessen Ergebnisse, die denen anderer AO_2-Systeme näher kommen. Es wird ein kooperativer Migrationsmechanismus von Sauerstoff- und Hafniumleerstellen vorgeschlagen. Die Leitfähigkeit von m-HfO_2 wurde mit Hilfe von Sauerstoffpartialdruckabhängigen Hochtemperatur-Gleichgewichtsleitfähigkeitsmessungen untersucht. Unter reduzierenden Bedingungen wurde festgestellt, dass die Gesamtleitfähigkeit mit dem Sauerstoffpartialdruck zunimmt. Numerische defektchemische Berechnungen zeigten, dass wahrscheinlich einfach positiv geladene Sauerstoffleerstellen für dieses Verhalten verantwortlich sind. In mittleren Sauerstoffpartialdruck-Regimen dominierte die ionische Leitfähigkeit. Unter oxidierenden Bedingungen stieg die Gesamtleitfähigkeit mit dem Sauerstoffpartialdruck aufgrund von Elektronenlöchern an.

Acknowledgements

This work would not exist without the help and support of Prof. Dr. Roger A. De Souza. His profound knowledge and understanding, his infectious enthusiasm and boundless support in this field of work have allowed me to successfully complete my thesis and left a deep impression on my own understanding of science and work ethic. I am deeply grateful for his supervision and could not have imagined a better supervisor.

I also want to give special thanks to Prof. Dr. Manfred Martin for giving me the opportunity to work in his group and follow my interest in solid-state physical chemistry.

I am deeply grateful to Prof. Dr. Richard Dronskowski for his review of this thesis, to Prof. Dr. Ulli Englert for being a examiner during the examination and to Prof. Dr. Walter Richtering for chairing the examination committee.

My gratitude goes to Dr. Alexandra von der Heiden, Dr. Annalena Genreith-Schriever, Christiane Ader and Jacqueline Börgers for the wonderful atmosphere in our office and the mountains of tea and sweets.

I would also like to thank the bachelor students I helped supervise during my thesis for being such motivated and skilled students. Dennis, Kendra, Katrin and Alexander, I wish you the very best for your future.

For proofreading this thesis, I am very thankful to Dennis Kemp, Joe Kler, Jana Parras, Jacqueline Börgers and Christiane Ader.

Special thanks go to Dr. Christian Schwab for supervising my research project years ago and continuing to provide assistance and fruitful discussion, Dr. Andreas Falkenstein for his helpfulness and scientific experience and the entire 'SpaMi' group for providing copious amounts of feedback on my research and fun times coupled with pizza and beer or Shetland ponies.

To the entire group I direct my warmest thanks; it was you who endured my bad jokes and sometimes even laughed at them, always provided a way to divert my attention and I can call many of you not only colleagues but also friends. I will count myself lucky if I find new colleagues as great as you.

Lastly, I want to thank my family and friends for always providing the support and distraction I needed when things got difficult. All of you are the best and I am incredibly thankful for you. To my dad, thank you for everything.

Contents

1 Introduction   1

2 Theory   5

     2.1 Properties of HfO_2 5

     2.2 Point Defects 6

     2.3 Diffusion in Solids 11

         2.3.1 Diffusion Mechanisms of Point Defects 13

     2.4 Ionic Conductivity and Mobility 15

         2.4.1 Influence of an Electrical Field on Ion Transport 19

     2.5 Extended Defects 21

         2.5.1 Grain Boundaries 21

         2.5.2 Space-Charge Theory 24

3 Experimental and Computational Methods   29

     3.1 Secondary Ion Mass Spectrometry 29

     3.2 Others 33

     3.3 Density-Functional-Theory 34

     3.4 Molecular Dynamics 39

| 4 | Oxygen Diffusion in HfO_2 | 45 |

    4.1 Introduction 45

    4.2 Experimental Details 46

    4.3 Results 51

    4.4 Discussion 52

    4.5 Summary 56

| 5 | Cation Diffusion in HfO_2 | 57 |

    5.1 Introduction 57

    5.2 Methodology 60

        5.2.1 Experimental 60

        5.2.2 Computational 64

    5.3 Results 66

        5.3.1 Experimental 66

        5.3.2 Computational 69

    5.4 Discussion 76

        5.4.1 Bulk diffusion 76

        5.4.2 Molecular Dynamics Simulations 78

        5.4.3 Grain-Boundary Diffusion 81

    5.5 Summary 83

| 6 | Electrical Properties and Defect Structure of HfO_2 | 85 |

    6.1 Introduction 85

        6.1.1 Acceptor-doped AO_2 Oxides 87

	6.2	Experimental Details	89
	6.3	Results	89
	6.4	Discussion	92
		6.4.1 Conductivity Measurements	92
	6.5	Summary	100
7	**Conclusion**		**101**
	Bibliography		**105**

Chapter 1

Introduction

Humanity strives to move forward, improve existing technologies and invent new ones; make them more powerful, more sustainable and, ideally, improve life for everyone. Therefore, interest in research on functional materials, proposing exciting new possibilities for many modern applications, is at an all-time high. One driving force of this ongoing process of adapting and improving existing materials is the miniaturisation of existing technologies. This applies specifically to energy storage or memory devices, known to most consumers for being important building blocks of smartphones. Manufacturers face challenges as current devices are reaching their size limits and further miniaturisation becomes seemingly impossible. Silica (SiO_2), one of the premier materials for dielectric gate devices, is characterised by high tunnelling currents once its layers become too thin and a subsequent loss of current in the device. This point has been reached, and it represents a roadblock for existing manufacturing techniques. This leads to the introduction of either new manufacturing processes, the search for new materials, or the improvement of current materials. The electronic properties of both old and new materials thus constitute an important focal point of research.

Since changing fabrication processes is an expensive project, consumers and manufacturers alike would prefer new and better suited materials, that do not require significant changes in process. Hafnia (HfO_2) is one such example, with the capability of downsizing electronic parts even further without worrying about tunnel currents. Interest in hafnia has subsequently grown dramatically in the last two decades. Due to its high dielectric permittivity, wide band gap and chemical and electronic compatibility with silicon, HfO_2 is gradually replacing SiO_2 in silicon-based semiconductor devices.[1,2] At the same time, research on defect transport in HfO_2 becomes more prominent because of role defects play in degradation processes, such as interdiffusion, kinetic demixing or creep. But not only is HfO_2 an important player in the realm of miniaturisation, it has various different applications. Forcing it to adopt a particular orthorhombic structure results in the appearance of ferroelectricity.[3] It is interesting for researchers working on new memory devices, such as Resistive Random Access Memories (ReRAMs),[4–7] which can be reversibly switched between high and low resistance states by means of suitable applied voltages. These so-called memristors use metal–insulator–metal structures and HfO_2 shows memristive behaviour when employed as the insulator.[8–18] While research on this topic is not restricted to HfO_2, it is one of the premier candidates for new memory devices. Oxide-ion transport in HfO_2, while neglected for many years, has thus become a point of interest, as it is relevant to understanding these resistive switching memory devices.

HfO_2 is employed as an amorphous solid when used in dielectric devices. While the thermodynamically stable structure at standard conditions is the cubic fluorite structure for most AO_2 oxides, it is the monoclinic phase for HfO_2. As a result, the transport of ions in non-cubic materials is also of fundamental interest. Furthermore, while a profound understanding of the conductive behaviour of HfO_2 is paramount to the research on memristors, previous studies focusing on the conductivity of HfO_2 fell short when it came to the conductivity in reducing conditions. However, for metal–insulator–metal structures, amorphous HfO_2 usually assumes a reduced form due to an applied current and hafnium metal as the metal layer.

Ion transport in crystalline solids generally takes place in a sublattice of mobile ions, and one or more sublattices of 'immobile' ions. The latter form a stable framework through which the mobile ions can migrate, usually with the help of defects. In reality, however, no ion is completely immobile in a crystalline solid: the 'immobile' ions are merely considerably less mobile than the mobile ions. If all ions were significantly mobile, the stable phase would not be solid. While modern analysis methods allow for much more in-depth investigation of ion transport in solids compared to earlier research, transport of 'immobile' ions is still experimentally challenging due to the need for high temperatures and long diffusion times to obtain diffusion data of sufficient quality. To alleviate these issues, computational studies can be helpful by providing complete control over the simulated system. Static simulations based on Density-Functional-Theory (DFT) or Empirical Pair Potentials (EPP) can provide all manner of defect generation and migration energies, also giving the user complete control over the number of defects. Alternatively, Molecular Dynamics (MD) allow direct simulations of ion transport as a function of temperature and subsequent extracting of diffusion coefficients and ion mobilities.

So far, only the fundamental motivation has been presented. The aim of this thesis, then, is to further the neglected understanding of ion transport in monoclinic HfO_2 (m-HfO_2) in order to improve future electronic devices.

To that end, first, oxygen diffusion in m-HfO_2 is investigated by fabricating dense bulk samples and performing oxygen exchange experiments and Secondary Ion Mass Spectrometry (SIMS) analysis. The results are compared with computational results from literature. Finite-Element-Method (FEM) simulations are needed to study the SIMS results and obtain diffusion coefficients and an activation enthalpy of diffusion.

Secondly, cation diffusion in m-HfO_2 is investigated, with zirconium serving as the cation of choice due to its chemical similarity with hafnium. Thin film samples are used instead of bulk ceramic samples because of findings from the oxygen diffusion experiments. SIMS and FEM simulations are once again used to obtain diffusivities and activation enthalpies of diffusion. The possibility of grain-boundary diffusion is also discussed. For comparison, DFT calculations are employed to obtain migration barriers. Additionally, the mobility of hafnium cations is probed with MD simulations combined with an applied electrical field to enhance cation migration.

Third, the conductivity of bulk m-HfO_2 is measured and the results are compared with diffusion data and literature values for the conductivity. Possible defect reactions responsible for the obtained conductivity curves are discussed.

Before the results of these three fundamental topics are presented in chapters 4, 5 and 6, chapter 2 explains the basic concepts which most findings in this thesis build upon. Chapter 3 then introduces the employed methods and a summary of the conclusions follows at the end in chapter 7.

Chapter 2

Theory

2.1 Properties of HfO$_2$

HfO$_2$ is a member of the AO_2 ($A^{IV}O_2^{-II}$) fluorite-type oxides. At room temperature and atmospheric pressure it crystallises in the monoclinic baddeleyite structure (m-HfO$_2$, space group P2$_1$/c), an off-white solid. It has a density of 9.68 g/cm^3 and the lattice parameters were found to be $a = 5.118$ Å, $b = 5.186$ Å and $c = 5.284$ Å, with an angle $\beta = 99.35°$. Phase transitions to a tetragonal phase (t-HfO$_2$, space group P4$_2$/nmc) and a cubic phase (c-HfO$_2$, space group $Fm\bar{3}m$) occur at ca. 1973 K and ca. 2773 K, respectively. This thesis focuses on the monoclinic and cubic phases, but for the sake of completeness it is noted that two orthorhombic phases exist. The first phase occurs at 4 to 14.5 GPa and below 1523 to 1673 K. The second phase occurs at 14.5 to 21 GPa and below 2073 K.[19]

In the cubic phase, hafnium cations are eightfold coordinated by oxygen anions in a cubic polyhedron. Each hafnium ion has twelve hafnium ions as next-nearest neighbours. The oxygen ions are tetrahedrally coordinated by four hafnium ions and have six oxygen ions as next-nearest neighbours. Both oxygen and hafnium have a single crystallographic site each and the unit cell consists of four formula units and twelve ions. The monoclinic phase is a distorted version of the cubic fluorite-type phase. While the number of next-nearest hafnium neighbours stays the same for the hafnium ions and hafnium remains on a single crystallographic site, each hafnium ion is now coordinated by seven oxygen ions. Oxygen has two

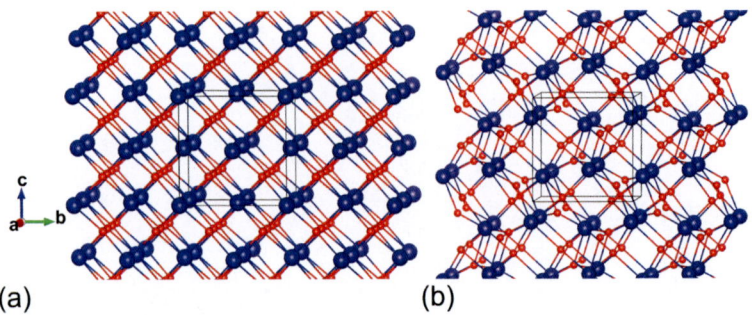

Figure 2.1: Hafnium ions are shown in blue, oxide ions in red. Size of the ions does not correspond to the real ionic size. Connecting lines between ions only indicate the coordinating ions. (a) Atomic arrangement in structure of c-HfO$_2$; (b) Atomic arrangement in structure of m-HfO$_2$.

crystallographic sites and is either threefold or fourfold coordinated by hafnium ions. The number of formula units and atoms in a unit cell remains the same as in the cubic structure. Fig. 2.1(a) shows the atomic arrangement in the cubic structure; Fig. 2.1(b) shows the atomic arrangement in the monoclinic structure.

HfO$_2$ exhibits many interesting properties, *i.e.* ferroelectricity, presumably originating in the orthorhombic phases,[3] or its high relative dielectric permittivity, especially important for the semi-conductor industry. It is different for each phase (*ca.* 15 for m-HfO$_2$, 31 for c-HfO$_2$, 126 for t-HfO$_2$ and 25 for a-HfO$_2$),[4,20,21] but all phases show a high dielectric constant compared to SiO$_2$ (*ca.* 4).[4] Because of this, HfO$_2$ is called a 'high-κ' dielectric with excellent electronic properties and applications in dielectric materials. The amorphous phase (a-HfO$_2$) is of the main focus for many dielectric applications, but not included in the scope of this work.

2.2 Point Defects

A perfect crystal does not exist. Crystalline solids always contain defects. These are not flaws, but cornerstones of the properties of the material. The defect chemistry strongly affects the electrical, optical or magnetic properties of solids. The smallest kind of defects are the point defects, which are limited in extent to one

2.2 Point Defects

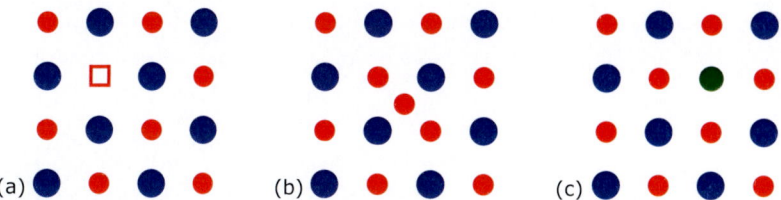

Figure 2.2: Various types of point defects. Cations shown in blue, anions in red and impurity ions or dopants are green. Type of ion does not matter for the definitions of these point defects. (a) Vacancy; (b) Interstitial ion; (c) Dopant or impurity ion.

single lattice site. These zero-dimensional defects can be vacancies, essentially empty sites in the lattice, and interstitial ions, occupying usually vacant sites. Another type of point defects are dopants, impurity ions foreign to the regular lattice. Fig. 2.2 shows the different types of typical point defects.

To discuss these defects, the Kröger–Vink notation is used as a way to quickly identify the type of defect.[22] It is written as

$$M_a^{q_{\text{rel}}}$$

where the defect species M is to be replaced by the chemical element of the described ion (or V for vacancy, e for an isolated electron and h for an electron hole). The subscript a denotes the site where the defect is located, given by either the chemical element of the ion usually located at this site or i for an interstitial site. The defect charge q_{rel} relative to the ion usually located at the site is written in superscript, with a prime / indicating a negative relative charge and a bullet point • indicating a positive relative charge. × indicates a relative charge of zero. An oxygen vacancy, for example, is written as $V_O^{\bullet\bullet}$, a hafnium ion on an interstitial position is $Hf_i^{\bullet\bullet\bullet\bullet}$. In the same fashion, a zirconium ion on a hafnium site in HfO_2 is $Zr_{Hf}^{\times}$.

To describe the influence of defects in solids, one has to be able to quantify the number of defects in a specific sample at certain conditions. This is by no means an easy feat, but generally a statistical problem. The concentration c_i of a particle i in a homogeneous solid is defined by a Boltzmann term. For dilute and non-interacting particles or defects c_i is equal to the number of possible lattice sites per volume N_i times the Boltzmann term containing the free energy of formation $\Delta G_{f,i}$ of the respective particle i, the temperature T and the Boltzmann constant k_B:

$$c_i = N_i \exp\left(-\frac{\Delta G_{f,i}}{k_B T}\right). \tag{2.1}$$

In solids, the unitless site fraction $n_i = \frac{c_i}{N_i}$ is sometimes used instead of the concentration c_i in units of cm^{-3}. Square brackets around element symbols can refer to either the concentrations or the site fractions. In this chapter, square brackets will indicate the use of site fractions.

A defect can either be intrinsic or extrinsic. Intrinsic defects are formed by reactions between species native to the system, meaning that intrinsic defects always exist in any given system. Extrinsic defects are introduced to the system through dopants. When looking at reactions between defects or the forming of new defects, the electroneutrality condition of the system is of utmost importance. The sum of the relative charge of **all** defects has to be zero. This becomes evident when looking at some typical intrinsic defect formation reactions in oxide compounds (with A^{4+} as any four valent cation species and O^{2-} as anion species, but the basic principle is applicable to all cation and anion species):

1. **Schottky Disorder**
   A Schottky disorder describes the formation of a surface species (with the same composition as the crystal) while creating both anion and cation vacancies in stoichiometric ratio in the crystal lattice:

   $$A_A^\times + 2O_O^\times \rightleftharpoons V_A'''' + 2V_O^{\bullet\bullet} + AO_2. \tag{2.2}$$

   The law of mass action takes the following form, with ΔH_{Sch} as the enthalpy and ΔS_{Sch} as the entropy of formation of a Schottky defect. The site fraction of the AO_2 species is omitted since it is unity:

   $$K_{\text{Sch}} = \frac{[V_A''''][V_O^{\bullet\bullet}]^2}{[A_A^\times][O_O^\times]^2} = \exp\left(-\frac{\Delta H_{\text{Sch}}}{k_B T}\right)\exp\left(\frac{\Delta S_{\text{Sch}}}{k_B}\right). \tag{2.3}$$

2. Frenkel Disorder

A Frenkel disorder describes the simultaneous formation of a cation interstitial and a cation vacancy. To that end, a cation on a regular crystallographic site leaves that site and occupies an empty interstitial site:

$$A_A^\times + V_i^\times \rightleftharpoons V_A'''' + A_i^{\bullet\bullet\bullet\bullet}. \tag{2.4}$$

The law of mass action takes the following form, with ΔH_F as the enthalpy and ΔS_F as the entropy of formation of a Frenkel defect:

$$K_\text{F} = \frac{[A_i^{\bullet\bullet\bullet\bullet}][V_A'''']}{[A_A^\times][V_i^\times]} = \exp\left(-\frac{\Delta H_\text{F}}{k_\text{B} T}\right) \exp\left(\frac{\Delta S_\text{F}}{k_\text{B}}\right). \tag{2.5}$$

3. Anti-Frenkel Disorder

An anti-Frenkel disorder describes the simultaneous formation of an anion interstitial and an anion vacancy. To that end, an anion on a regular crystallographic site leaves that site and occupies an empty interstitial site:

$$O_O^\times + V_i^\times \rightleftharpoons V_O^{\bullet\bullet} + O_i''. \tag{2.6}$$

The law of mass action takes the following form, with ΔH_aF as the enthalpy and ΔS_aF as the entropy of formation of an anti-Frenkel defect:

$$K_\text{aF} = \frac{[O_i''][V_O^{\bullet\bullet}]}{[O_O^\times][V_i^\times]} = \exp\left(-\frac{\Delta H_\text{aF}}{k_\text{B} T}\right) \exp\left(\frac{\Delta S_\text{aF}}{k_\text{B}}\right). \tag{2.7}$$

4. Reduction and Oxidation

The concentration of oxygen vacancies is dependent on the oxygen activity aO_2. As long as the surface reaction is fast enough, any given system tries to adjust to the atmosphere by incorporating or removing oxygen ions from surrounding oxygen molecules and the amount of oxygen vacancies subsequently increases or decreases. Electroneutrality is maintained through other cations accepting or donating electrons:

$$O_O^\times + 2A_A^\times \rightleftharpoons V_O^{\bullet\bullet} + \frac{1}{2}O_2 + 2A_A'. \tag{2.8}$$

The law of mass action for the reduction process takes the following form, with ΔH_{Red} as the enthalpy and ΔS_{Red} as the entropy of formation of a Reduction:

$$K_{\text{Red}} = \frac{[V_O^{\bullet\bullet}]\, aO_2^{\frac{1}{2}}\, [A'_A]^2}{[O_O^\times]\, [A_A^\times]^2} = \exp\left(-\frac{\Delta H_{\text{Red}}}{k_B T}\right) \exp\left(\frac{\Delta S_{\text{Red}}}{k_B}\right). \quad (2.9)$$

The law of mass action for the oxidation process is defined analogous in terms of holes instead of electrons.

For these intrinsic defects the sum of the relative charges on each side of the equations above is zero. That means the electroneutrality condition is met. For extrinsic defects stemming from dopants, each side of the electroneutrality condition contains either all positively or negatively charged species and the absolute value of charges on both sides is equal:

$$[Ac'_A] + 2\,[O''_i] + 4\,[V''''_A] = 2\,[V_O^{\bullet\bullet}] + 4\,[A_i^{\bullet\bullet\bullet\bullet}] + [D_A^\bullet]. \quad (2.10)$$

Dopants are differentiated between donors (D), which carry one more valence electron than the species native to the lattice site and become positively charged ($D^\bullet$) upon doping, and acceptors (Ac), which carry one less valence electron than the native species and become negatively charged (Ac') upon doping. The concentrations of all defects are connected according to Eq. 2.10 and it is not possible to change the concentration of only one defect species.

Point defects influence a plethora of properties of solid functional materials and they are also needed for ion transport in crystalline solids. How this transport is related to thermodynamic quantities is described by the concepts of diffusivity and conductivity.

2.3 Diffusion in Solids

Diffusion in solids refers to the long-range displacement of ions in the atomic lattice. For diffusion to take place, a gradient is required. **Chemical diffusion** of a component takes place as a result of a gradient in chemical potential of the relevant component. This difference generally exists between two different systems connected via an interface or due to different chemical compositions in a single sample. To bring the system into thermodynamical equilibrium, diffusion takes place.

In the absence of a gradient, only **self diffusion** according to random-walk theory can take place. While observing self diffusion is impossible, **tracer diffusion** can be observed and is approximately identical to self diffusion, as long as the tracer is chemically identical to the surrounding matrix. For an infinitely diluted system (meaning that c_{def} is small and $c_{ion} \approx N_{def}$), the proportionality of tracer diffusion and self diffusion can be expressed as follows:

$$D^* = f^*D = f^*D_{def}n_{def} = f^*D_{def}\frac{c_{def}}{N_{def}}. \tag{2.11}$$

The tracer diffusion coefficient D^* differs from the self-diffusion coefficient D only via the tracer correlation factor f^*, which is dependent on the lattice geometry and the mechanism of diffusion. This factor is needed since tracer diffusion is not completely random, but correlated. For example, for anion diffusion in cubic fluorite structures $f^* = 0.653$.[23] Furthermore, Eq. 2.11 and the tracer diffusion coefficient allow the determination of the diffusion coefficient of a defect species, as long as the defect fraction n_{def} is known.

On a macroscopic scale, Fick's laws of diffusion can generally be used to quantitatively describe diffusion:

$$J = -D\frac{\partial c}{\partial x}. \tag{2.12}$$

Fick's first law illustrates how the flux J, a number of particles that traverse a specified area in a specified time interval, is proportional to the negative concentration gradient in a specified direction x, where the proportionality factor is the diffusion coefficient D. The continuity equation describes the change in concentration over time in relation to the positional change of flux:

$$\frac{\partial c}{\partial t} = -\frac{\partial J}{\partial x}. \tag{2.13}$$

By combining Eq. 2.13 and Eq. 2.12 Fick's second law is obtained, which describes the change in concentration due to the diffusion process (if D is independent from c and X):

$$\frac{\partial c}{\partial t} = D\frac{\partial^2 c}{\partial x^2}. \tag{2.14}$$

This differential equation can be solved for specific boundary conditions and the solutions are available in the literature for many common experimental setups and study cases.[24] One such example, that is also used in this thesis, is the solution of the diffusion equation for a semi-infinite, homogeneous medium with surface limitations and a constant diffusion source. The following boundary conditions apply (using tracer fractions n^* instead of concentrations c^*):

1. At time $t = 0$ the tracer fraction n^* in the samples is equal to the background fraction n^*_{bg} of the tracer in the sample:

$$n^*(x, t = 0) = n^*_{\text{bg}}. \tag{2.15}$$

2. At all times, at a point in the sample infinitely far away from the surface, n^* is also equal to n^*_{bg}:

$$n^*(x = \infty, t) = n^*_{\text{bg}}. \tag{2.16}$$

3. The difference between tracer concentration at the surface n^*_{s} and in the gaseous diffusion source n^*_{gas} is proportional to the particle flux at the surface. The proportionality constant k^* is called surface exchange coefficient:

$$-D^*\left(\frac{\partial n^*}{\partial x}\right)_{x=0} = k^*\left(n^*_{\text{gas}} - n^*_{\text{s}}(t)\right). \tag{2.17}$$

2.3 Diffusion in Solids

Using these boundary conditions, a solution of Eq. 2.14 is obtained:[24, p. 36]

$$n^*(x) = \text{erfc}\left(\frac{x}{2\sqrt{D^*t}}\right) - \exp\left(\left(\frac{k^*}{D^*}\right)x + \left(\frac{k^*}{D^*}\right)^2 D^*t\right)$$
$$\text{erfc}\left(\frac{x}{2\sqrt{D^*t}} + \frac{k^*}{D^*}\sqrt{D^*t}\right).$$
(2.18)

2.3.1 Diffusion Mechanisms of Point Defects

Three main kinds of diffusion mechanisms can be observed for ions in solids, all requiring point defects. The jumps depicted here are not necessarily linear, depending on the crystal structure.

First is the **vacancy mechanism** (Fig. 2.3) — an ion migrates from a regular crystallographic site to a vacant site in its neighbourhood, so that ion and vacancy switch places. By repeating this process through the entire crystal ion diffusion takes place.

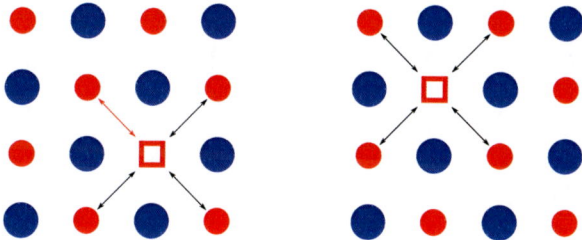

Figure 2.3: Cations are drawn in blue, anions are drawn in red. Schematic example of the vacancy mechanism. An ion migrates from a regular crystallographic site to a vacant site. Red arrow indicates the ion jump that takes place, black arrows indicate the other possibilities.

Analogous to the vacancy mechanism exists the **interstitial mechanism** (Fig. 2.4) in which an ion migrates from an interstitial crystallographic site to a vacant interstitial site in its neighbourhood.

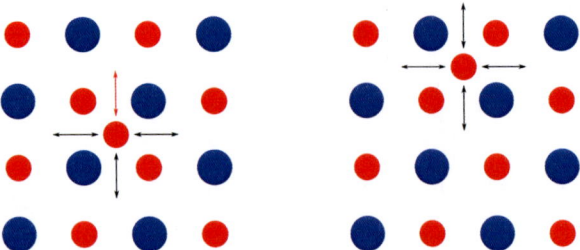

Figure 2.4: Schematic example of the interstitial mechanism. An ion migrates from an interstitial crystallographic site to a vacant interstitial site.

Lastly, there is the **interstitialcy mechanism** (Fig. 2.5) — an ion migrates from an interstitial crystallographic site to a neighbouring regular site and the displaced ion moves to a vacant neighbouring interstitial site instead.

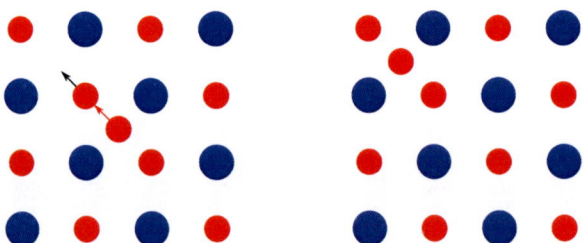

Figure 2.5: Schematic example of the interstitialcy mechanism. An ion migrates from an interstitial crystallographic site to a neighbouring regular site and the displaced ion moves to a vacant neighbouring interstitial site. Red arrow indicates the ion jump that takes place first, black arrow indicates the subsequent jump.

In c-HfO$_2$ a vacancy mechanism is assumed for cation and anion diffusion. In c-AO_2 fluorite-type oxides, cations jump along the $\langle 110 \rangle$ direction to vacant sites.[25–28] The high degree of symmetry in c-AO_2 leads to all twelve possible jumps being symmetry equivalent. Other migration paths are possible but much less likely due to higher migration barriers. Oxygen anions each have six neighbouring, equidistant oxygen sites, which means that all jumps in $\langle 100 \rangle$ direction are equally favoured. For m-HfO$_2$ the diffusion mechanism is more complex due to the monoclinic distortion. Even though the number of nearest-neighbour cations remains the same for hafnium, the 12 possible jumps are no longer symmetry equivalent, in fact, 8 different jumps are now possible. The favoured jump direction of the cations will be discussed later on. Oxygen diffusion in m-HfO$_2$ takes place via jumps with large $\langle 110 \rangle$ components,[29] but a single jump is not sufficient in m-HfO$_2$ for an oxygen ion to traverse from one end of the cell to the other end, making a combination of different jumps necessary.

2.4 Ionic Conductivity and Mobility

While Fick's laws can be applied to describe ion diffusion in solids on a macroscopic scale, the microscopic scale is also of relevance for our understanding of diffusion. Point defects do not appear in Fick's model but are necessary for diffusion in solids since vacant sites are needed for movement. Strictly speaking, without neighbouring vacant lattice sites, an ion remains trapped on its own lattice site (barring lattice vibrations). When looking at diffusion in terms of singular moving particles in a three-dimensional space, Einstein and Smoluchowski suggested the following:

$$\langle \mathbf{r}^2 \rangle = 6Dt. \tag{2.19}$$

$\langle \mathbf{r}^2 \rangle$ denotes the mean squared displacement (msd):

$$\langle \mathbf{r}^2 \rangle = \langle |\mathbf{r}(t) - \mathbf{r}(t_0)| \rangle. \tag{2.20}$$

The msd describes the squared displacement from each particles point of origin at a specified point in time (generally averaged over all particles). The factor 6 stems from the three-dimensional case, the particle has 2 possible jump directions in each dimension. Assuming the point defect is infinitely diluted, the jumps are uncorrelated.

For uncorrelated diffusion processes and known jump distances a_{jump}, random-walk theory states the following:

$$\langle \mathbf{r}^2 \rangle = n_{\text{jump}} a_{\text{jump}}^2. \quad (2.21)$$

n_{jump} is the number of jumps. The number of jumps is related to the number of nearest neighbours N_N and the jump rate Γ of the defect:

$$N_N \Gamma = \frac{n_{\text{jump}}}{t}. \quad (2.22)$$

The jump rate is temperature dependent and follows an Arrhenius-type behaviour:

$$\Gamma = \nu_0 \exp\left(-\frac{\Delta G_{\text{mig}}}{k_B T}\right). \quad (2.23)$$

ΔG_{mig} is the necessary free energy of migration the defect has to overcome to execute the jump in question. The attempt frequency ν_0 denotes the frequency at which a jump is attempted and is generally similar in value to the Debye frequency. Combining Eq. 2.19, Eq. 2.21 and Eq. 2.22 yields an expression for the relation of the diffusion coefficient to the jump rate and length:

$$D = \frac{1}{6} a_{\text{jump}}^2 N_N \Gamma. \quad (2.24)$$

Therefore, by combining Eq. 2.24 and Eq. 2.23 an expression for the diffusion coefficient of a point defect is obtained:

$$D = \frac{1}{6} a_{\text{jump}}^2 N_N \nu_0 \exp\left(-\frac{\Delta G_{\text{mig}}}{k_B T}\right). \quad (2.25)$$

2.4 Ionic Conductivity and Mobility

It is often more suitable to use the activation enthalpy of migration ΔH_{mig} instead of the free energy ΔG_{mig}. Eq. 2.26 describes the relation between free energy and enthalpy by introducing the entropy of migration ΔS_{mig}:

$$\Delta G_{mig} = \Delta H_{mig} - T\Delta S_{mig}. \tag{2.26}$$

Expression 2.25 can then be simplified by introducing a pre-exponential factor D° containing the factors in front of the exponential term and the entropy term:

$$D = D^\circ \exp\left(-\frac{\Delta H_{mig}}{k_B T}\right). \tag{2.27}$$

The resulting eq. 2.27 is simple to linearise, allowing the analytical determination of migration enthalpies by measuring or simulating temperature dependent diffusion coefficients (assuming diluted solution):

$$\ln D = \ln D^\circ - \frac{1}{T}\frac{\Delta H_{mig}}{k_B}. \tag{2.28}$$

The diffusion coefficients of ions and their respective vacancies are connected. The number of vacancy jumps must be equal to the number of associated ion jumps. For an ion i and its vacancy v follows:

$$D_i = \frac{c_v}{c_i} D_v. \tag{2.29}$$

The diffusivity of a particle is dependent on its mobility. For charged particles, such as ions, the mobility is influenced by external electrical fields. For small fields the following relation exists between diffusion coefficient D_i and mobility u_i, with $|z_i|$ as the charge number of the ion and e as the electron charge:

$$u_i = \frac{D_i |z_i| e}{k_B T}. \tag{2.30}$$

The mobility of an ion is one of two main components to the conductivity σ_i. The other component is the concentration c_i of conductors. For an infinitely diluted system, the conductivity of an ion conductor is defined as

$$\sigma_i = c_i |z_i| e u_i. \tag{2.31}$$

Using Eq. 2.25, 2.29, 2.30 and 2.31 an expression for the conductivity of an ion dependent on the diffusivity of said ion is obtained:

$$\sigma_i = \frac{|z_i|^2 e^2 a_{\text{jump}}^2 N_N \nu_0}{6 k_B T} c_v \exp\left(-\frac{\Delta G_{\text{mig}}}{k_B T}\right). \tag{2.32}$$

2.4 Ionic Conductivity and Mobility

2.4.1 Influence of an Electrical Field on Ion Transport

The migration enthalpy ΔH_{mig} describes the energy barrier a migrating particle has to overcome to jump to a different vacant lattice site at a constant pressure and temperature. It corresponds to the local maximum of the energy profile between the starting state and the final state of the jump (see also Eq. 2.23). Since ions are charged particles, an external electrical field has an effect on migration barriers and ion movement in general. This can be seen in Fig. 2.6.

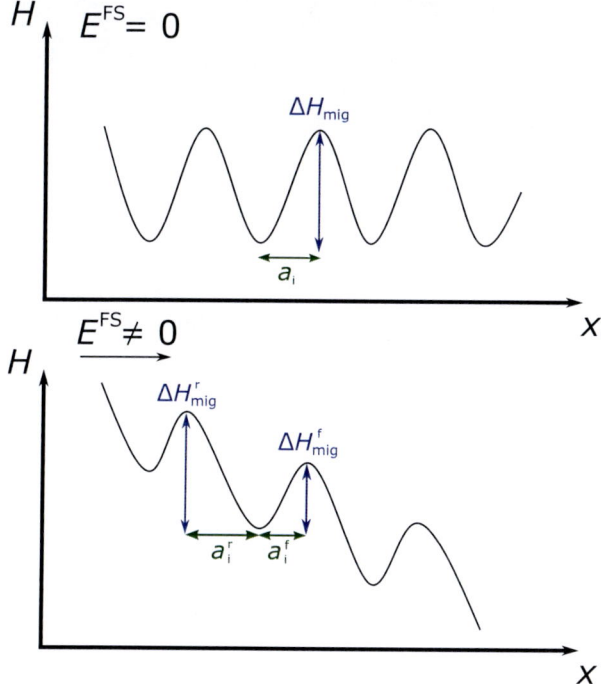

Figure 2.6: Influence of an electrical field E^{FS} on the migration barriers in a solid. For a non-zero field the migration barrier ΔH_{mig} is lowered in field direction ($\Delta H^{\text{f}}_{\text{mig}}$) and heightened against field direction ($\Delta H^{\text{r}}_{\text{mig}}$).

Verwey, Mott and Gurney were the first to explain the effects of the electrical field.[30,31] They found that the migration barrier of a migrating ion is lowered or increased under the influence of an external electrical field, depending on the applied field strength E^{FS}:

$$\Delta H_{\text{mig}}^{\text{f/r}} = \Delta H_{\text{mig}} \mp |z_i| \, eE^{\text{FS}} a_i. \qquad (2.33)$$

$\Delta H_{\text{mig}}^{\text{f}}$ denotes the lowered migration barrier in forward direction and $\Delta H_{\text{mig}}^{\text{r}}$ is the increased barrier in reverse direction. The distance between the initial location of the ion and the saddle-point configuration is denoted as a_i and is equal to $a_{\text{jump}}/2$ for linear jumps. The drift velocity v_d of the mobile ions can then be expressed as follows:

$$v_d = B \cdot \left[\exp\left(-\frac{\Delta H_{\text{mig}}^{\text{f}}}{k_B T}\right) - \exp\left(-\frac{\Delta H_{\text{mig}}^{\text{r}}}{k_B T}\right) \right], \qquad (2.34)$$

with $B = 2(1-n_i)\, a_i \nu_0 \exp\left(\frac{\Delta S_{\text{mig}}}{k_B}\right)$ and ΔS_{mig} as introduced in Eq. 2.26. Substituting Eq. 2.33 in Eq. 2.34 results in Eq. 2.35:

$$v_d = 2B \exp\left(-\frac{\Delta H_{\text{mig}}}{k_B T}\right) \sinh\left(\frac{|z_i|\, eE^{\text{FS}} a_i}{k_B T}\right). \qquad (2.35)$$

For small field strengths the term $\sinh(x)$ approximates to x, making $v_d \propto E^{\text{FS}}$ and the proportionality linear. As a result, the proportionality of drift velocity v_d and field strength E^{FS} is expressed by the ionic mobility $u_i = v_d/E^{\text{FS}}$. u_i is field-independent in this case. For large field strengths, however, an exponential relationship is found: $v_d \propto \exp\left(\frac{|z_i|eE^{\text{FS}} a_i}{k_B T}\right)$, making the mobility strongly field-dependent.

Genreith-Schriever and De Souza[32] have shown that the standard model of Verwey, Mott and Gurney fails to describe the ionic mobility in the high-field regime, for example in the cubic fluorite-type oxide CeO_2.

They developed a correction to the old model that imposes a linear field on the energy landscape:

$$H(x) = \frac{\Delta H_{\text{mig}}}{2} \cos\left(\frac{\pi x}{a_i}\right) - |z_i|\, eE^{\text{FS}} x. \tag{2.36}$$

The migration barrier $\Delta H_{\text{mig}}^{\text{f/r}}$ is analytically derived:

$$\begin{aligned}\Delta H_{\text{mig}}^{\text{f/r}} &= H\left(x_{\text{max}}^{\text{f/r}}\right) - H\left(x_{\text{min}}\right) \\ &= \Delta H_{\text{mig}} \left[\sqrt{1-\gamma^2} \mp \gamma\left(\frac{\pi}{2}\right) + \gamma \arcsin\gamma\right],\end{aligned} \tag{2.37}$$

with $\gamma = \frac{(2|z_i|eE^{\text{FS}} a_i)}{(\pi \Delta H_{\text{mig}})}$.

Using this model, the influence of an external electrical field on ion migration in fluorites like CeO_2 or HfO_2 can be accurately described.

2.5 Extended Defects

While point defects are unavoidable in solids, some extended defects may appear. These extended defects are always of non-zero dimensionality. Among them are dislocations (one-dimensional), grain boundaries and surfaces (two-dimensional), and pores (three-dimensional). Contrary to point defects, these extended defects are generally metastable and depend on the history of the solid. Still, they can significantly influence mechanical and electrical properties of materials and have a large impact on ion transport.

2.5.1 Grain Boundaries

Grain boundaries are two-dimensional interfaces between crystallites in polycrystalline solids. Their existence depends greatly on the thermal and mechanical history of the sample, as they occur on the interfaces of two differently oriented crystal grains. They disrupt the periodicity of the crystal and are generally defined by the angle between the two meeting grains, the lattice plane shared between the two grains and the tilt or rotation of the grain. Grain boundaries are reported to cause fast grain-boundary diffusion, in that the ratio of the activation enthalpies of grain-boundary diffusion (ΔH_{gb}) and bulk diffusion (ΔH_b) is often found to be $r = \Delta H_{gb}/\Delta H_b \approx 0.5$.[33–37] This is not necessarily the case, however, as the ratio of the activation enthalpies was also found to be $r = \Delta H_{gb}/\Delta H_b \leq 1$ for some AO_2 type oxides.[38,39] Furthermore, grain boundaries can introduce space-charge layers into polycrystals, capable of both increasing or hindering diffusion.[40–45]

The influence of grain boundaries on diffusion properties is often characterised by one of three kinetic diffusion regimes proposed by Harrison.[46]

Harrison type A: This type of diffusion regime applies when the diffusion coefficients in the bulk D_b and in the grain boundary D_{gb} are similar and diffusion takes place for a long time t at high temperatures T, while the grain size d_g is small. This results in the diffusion length $\sqrt{D_b t}$ being greater than the grain size (see Fig. 2.7(a)):

$$\sqrt{D_b t} \gg d_g \qquad (2.38)$$

It is usually not possible to differentiate between this type of grain-boundary diffusion and bulk diffusion, because the diffusion front is almost completely planar. As a result, only an effective diffusion coefficient D_{eff} is obtainable.

Harrison type C: The counterpart to the type A regime is the type C diffusion regime. It denotes a diffusion mainly via grain boundaries. Diffusion in the bulk is slow compared to the grain boundaries, so that almost no diffusing particles enter the bulk from the grain boundary. Furthermore, t is too short for significant bulk diffusion to have occurred. Fig. 2.7(b) shows that diffusion takes place only in the grain-boundary width ω.

$$\sqrt{D_b t} \ll \omega \qquad (2.39)$$

Only the grain-boundary diffusion coefficient D_{gb} can be experimentally determined.

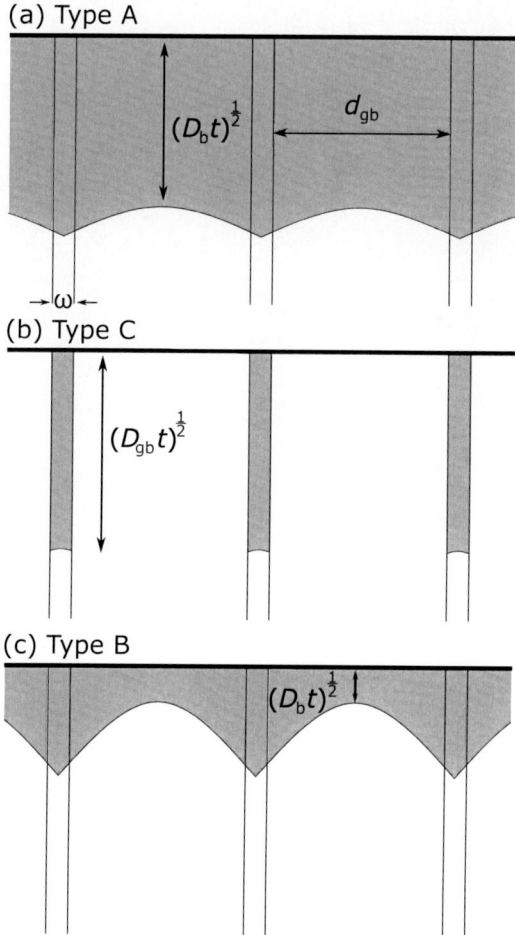

Figure 2.7: Classification of grain-boundary diffusion kinetics according to Harrison.[46] Type A diffusion kinetics do not allow differentiation between grain-boundary and bulk diffusion. Type C diffusion kinetics are defined by almost pure grain-boundary diffusion. Type B diffusion kinetics allow differentiating between grain-boundary and bulk diffusion.

Harrison type B: This type of diffusion regime is common in polycrystalline materials. It takes place when the diffusion length $\sqrt{D_b t}$ is greater than the grain-boundary width ω, but smaller than the grain size d_g:

$$\omega \ll \sqrt{D_b t} \ll d_g \tag{2.40}$$

The diffusing particles form a diffusion front throughout the bulk material, but are able to migrate further into the sample via grain boundaries (see Fig. 2.7(c)) and diffuse into the outer layers of grains from there. Grain-boundary diffusion and bulk diffusion is generally easy to differentiate for this diffusion regime, since both types of diffusion generate distinct diffusion profiles. While D_b can be described through solution of the diffusion equation, determining D_{gb} is more complicated, although possible. This is described in detail elsewhere.[47,48]

2.5.2 Space-Charge Theory

Equilibrium space-charge layers occur in ionic solids when point defects interact with extended defects like grain boundaries, where the periodic structure of the crystal is perturbed. This is described as a **core | space-charge layer** model — with the **core** referring to the centre of the extended defect where the structure is significantly altered compared to the bulk phase of the crystal. The **space-charge layer** describes the 3-dimensional region around the core extending into the bulk crystal. Diffusion properties in the space-charge layers might be different compared to the bulk of the crystal.[49,50] Fig. 2.8 depicts an exemplary case for an oxide with immobile acceptor cations which compensate mobile oxygen vacancies.

Since the defect chemistry in the core of the extended defect is generally significantly altered compared to the bulk, point defects in the core have a different standard chemical potential and subsequently a different defect formation energy than the point defects in the bulk of the material (Fig. 2.8(a)). For this example, the point defects are assumed to be oxygen vacancies. If the oxygen vacancies are mobile, they redistribute to accumulate in the core and deplete in the region next to the core (Fig. 2.8(b)). The vacancy-depleted regions form the space-charge layers, also called Mott–Schottky layers. Since acceptor cations are immobile compared to oxygen vacancies and do not significantly redistribute, the acceptor concen-

2.5 Extended Defects

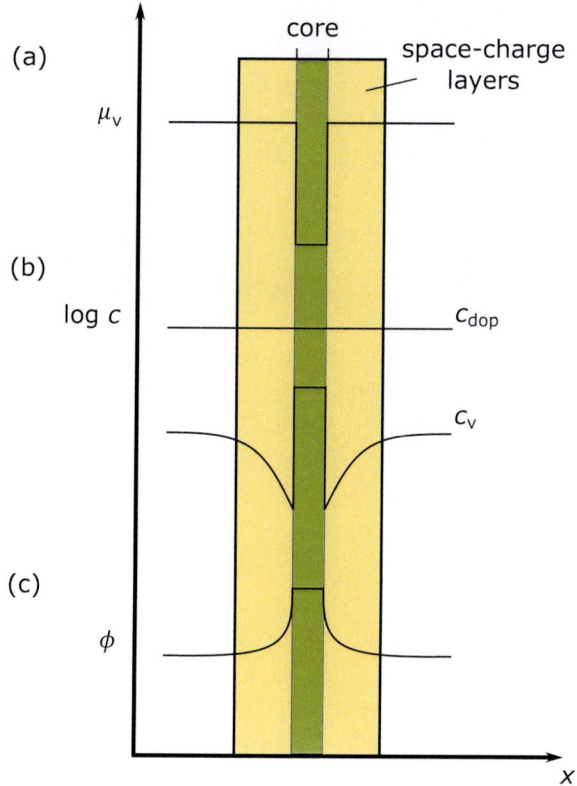

Figure 2.8: Characteristics of space-charge layers in an oxide with mobile oxygen vacancies and immobile acceptor ions. (a) Oxygen vacancy formation energy; (b) Oxygen vacancy concentration profile and dopant concentration profile; (c) Electrostatic potential.

tration is seen as constant. The accumulation of (positive) vacancies in the core and depletion in the space-charge layer then creates a positive electrostatic potential ϕ in the core (Fig. 2.8(c)). N.b. that the extent of the space-charge layers is much greater than the extent of the core. Since oxygen vacancies accumulate in the core, an increase in diffusivity of the respective species along the core is expected (ignoring the likely effect of the differing structure in the core). However, since the space-charge layers feature a depletion of oxygen vacancies and are generally much more extensive, diffusivity perpendicular to the extended defect is generally decreased, assuming that mobilities are homogeneous over the material. Ultimately, whether space-charge layers lead to an increase or a decrease in ion conductivity over the entire system is a non-trivial question dependent on many variables. It can, however, be solved numerically. To do this, an equation to describe the distribution of ions or defects and a connection to the potential ϕ is needed.

The electrochemical potential $\tilde{\mu}_i$ dependent on the electrostatic potential ϕ is defined as:

$$\tilde{\mu}_i = \mu_i^\circ + k_B T \ln n_i + z_i e \phi, \tag{2.41}$$

with μ_i° as the standard chemical potential, n_i as the site fraction of species i and $z_i e$ as the charge of species i. As long as μ_i° remains constant, the gradient of the electrochemical potential $\nabla \tilde{\mu}_i$ is then defined as:

$$\nabla \tilde{\mu}_i = k_B T \nabla \ln n_i + z_i e \nabla \phi, \tag{2.42}$$

In equilibrium the gradient in electrochemical potential is zero for both the ion species and the respective defect species. By then integrating Eq. 2.42, Eq. 2.43 is obtained (with $n_i = c_i/N_i$):

$$\ln\left(\frac{c_i(x)}{c_i(\infty)}\right) = \ln\left(\frac{c_{\text{def}}(x)}{c_{\text{def}}(\infty)}\right) + \frac{z_i e [\phi(x) - \phi(\infty)]}{k_B T}. \tag{2.43}$$

This equation describes the relation of the electrostatic potential and the concentration of species i. Both are position-dependent, with ∞ referring to the bulk phase of the system and x denoting the position in the space-charge layer. In case the difference in the electrostatic potential $\Delta \phi$ is positive and the defect concentration in the bulk $c_{\text{def}}(\infty)$ is small, the following approximate expression for the position-dependent defect concentration is obtained:

$$c_{\text{def}}(x) = c_{\text{def}}(\infty) \exp\left(\frac{-z_i e [\phi(x) - \phi(\infty)]}{k_B T}\right) \tag{2.44}$$

2.5 Extended Defects

The position-dependent electrostatic potential $\phi(x)$ can be obtained by solving the Poisson equation. This equation describes the relation of the electrostatic potential and the defect concentration:

$$\epsilon_0 \epsilon_r \nabla^2 \phi(x) = -\rho(x) = -\sum_i [ez_i c_i(x)]. \tag{2.45}$$

ϵ_r and ϵ_0 denote the relative dielectrical permittivity and vacuum permittivity. $\rho(x)$ is the space-charge density. As long as the defect species and its concentration are known, solving the Poisson equation is possible. N.b. that only the mobile defect species (here: oxide vacancies) are affected by the electrostatic potential, immobile defect species (here: singly charged acceptors) are not. Assuming that the electrostatic potential in the bulk is constant ($\nabla \phi(\infty) = 0$) and that the space-charge potential ϕ_0 at the interface $x = 0$ is defined as $\phi_0 = \phi(0) - \phi(\infty)$, one obtains:

$$\epsilon_0 \epsilon_r \nabla^2 \phi_0 = ec_{\text{def}} - 2ec_i(\infty) \exp\left(\frac{-2e[\phi(x) - \phi(\infty)]}{k_B T}\right) \tag{2.46}$$

This equation can be numerically solved to obtain a position-dependent distribution of defects or ions. Combining Eq. 2.46 with Eq. 2.11 gives the following expression for oxide ion diffusion through a space-charge layer:

$$D_i^*(x) = D_i^*(\infty) \exp\left(\frac{-2e[\phi(x) - \phi(\infty)]}{k_B T}\right) \tag{2.47}$$

Chapter 3

Experimental and Computational Methods

3.1 Secondary Ion Mass Spectrometry

Investigation of tracer diffusion in solids can be done by various methods, one of which is Secondary Ion Mass Spectrometry (SIMS). There are several types of SIMS machines which differ in the way they detect secondary ions, for example quadrupole SIMS, magnetic sector SIMS and the type used in this work, Time-of-Flight SIMS (ToF-SIMS). It separates secondary ions by their mass-to-charge ratio via a time of flight measurement. To analyse a specific species of ions, first primary ions are shot in short pulses at the sample surface in an ultra-high vacuum (to prevent surface contaminations) using an ion gun. Penetration depth of the primary ions depends on mass, energy, impact angle and pulse time. Typical primary ions for ToF-SIMS are Ga^+ or Bi_n^+ clusters, accelerated with energies of 15 to 30 keV. These primary ions cause collision cascades by colliding with atoms in the sample. The region of the sample that is changed by the collision cascade is called 'altered layer'. Fig. 3.1(a) displays a schematic collision cascade. Once particles reach the surface, they are able to leave the sample and are emitted. Most of the emitted particles are neutral and are not analysed any further, some are charged and are drawn to an extraction electrode. 3.1(b) shows a sketch of a ToF-SIMS machine. The ions drawn to the extraction electrode are the secondary ions and they are then sent into the flight tube, where they are separated by their mass-to-

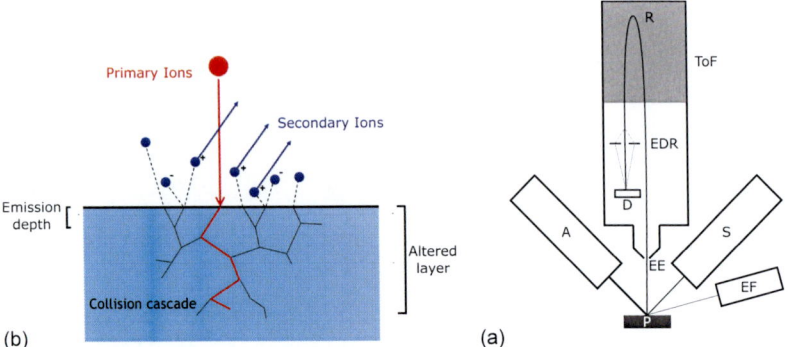

Figure 3.1: (a) Sketch of collision cascade and altered layer. Adapted after ref. 51; (b) Example sketch of a ToF-SIMS machine. Shown are the sample (P), primary ion gun (A), sputter gun (S), electron gun (EF), extraction electrode (EE), flight tube (ToF), reflectron (R), extended dynamic range (EDR) detector and the normal detector (D).

charge ratio and then detected. The relation of flight time t to the mass-to-charge ratio $\frac{m}{q}$ is described by Eq. 3.1:

$$t = \sqrt{\frac{l^2}{2U}\frac{m}{q}}, \qquad (3.1)$$

with l as the length of the flight path and U as the voltage applied at the extraction electrode. Identical secondary ions are often ejected at different velocities, causing them to arrive at the detector after different flight times. The reflectron compensates these differences in speed so that identical secondary ions arrive at the detector in the same time frame. Faster ions penetrate the electrical field generated by the reflectron deeper than slow ions, causing the fast ions to take a longer flight path and arriving at the same time as the slow ones. Secondary ions exhibiting a very high intensity can overload the detector and exceed the maximum count rate of ions, resulting in a loss of linear detector response (where the detected counts of a secondary-ion species increases linearly with the primary ion current). To alleviate this problem an Extended Dynamic Range (EDR) detector can be used. The basic principle is simple, the secondary ions beam coming from the sample and shortly before hitting the detector can be deflected into one of two attenuation lattices where either 90 % or 99 % of ions are absorbed. The remaining secondary ions are then accelerated via an electrical field so that their time of flight is not changed by the deflection. This way, the intensity of secondary ions can be selectively attenuated and because the attenuation factors of the lattices are known, the true intensity of the attenuated secondary ions can be calculated. N.b. that attenuated secondary-ion species show more signal noise, meaning that EDR should only be carefully used. There also exists an alternate, somewhat deprecated solution to the overloading problem, and that is the burst mode. It utilises separating a single pulse of primary ions into several shorter pulses. Several, much lower intensity pulses subsequently arrive at the detector. The downside to this method is the worse mass resolution due to the larger pulse width of the primary ions needed. The single-pulse method is also called bunched mode.[52]

To obtain depth resolution, another ion source is used as a sputter gun. It shoots positively charged ions at the sample surface, similar to the primary ion gun. Typical sputter ions for ToF-SIMS are Cs^+, O_2^+ or O_n^+ clusters and they are accelerated with low energies between 0.25 and 2 keV. The sputter ions are used to abrade the surface and uncover new sample layers. By switching between primary ion gun and sputter gun the ToF-SIMS machine alternates between analysis of secondary ions and removing surface layers, thus obtaining depth resolution. The apparent downside to SIMS is the destructive nature, since ion bombardment changes the sample irreversibly.

Samples with low conductivity are prone to accumulate charge from the primary and sputter ions around the analysed spot. This leads to a plethora of undesirable effects, such as degradation of mass resolution and at worst primary ions are no longer able to reach the sample surface. For this reason an electron flood gun is available to flood the sample with electrons and equilibrate the charge.

Upon arrival of the secondary-ion species A at the detector, the mass spectrometer counts the ions and measures an intensity. Several factors contribute to this intensity I_A^{SI}, which are explained in the so-called SIMS-equation:

$$I_A^{SI} = I^{PI} \gamma \alpha_A^{+/-} \eta_A \theta_A c_A. \tag{3.2}$$

The different parameters are, in order, the intensity of the primary ions I^{PI}; the sputter yield γ, which describes the amount of fragments emitted from the sample per primary ion; the ionisation probability $\alpha_A^{+/-}$, depicting the amount of charged fragments per emitted fragment; the detection factor η_A, describing the amount of detected ions A per emitted ion; the isotopic ratio θ_A and the atomic fraction c_A. Many of these parameters are dependent on not only the investigated species, but also on the sample matrix. The sputter yield γ is dependent on the type, energy and angle of the primary ions, while the ionisation probability $\alpha_A^{+/-}$ can be influenced by choosing different sputter ions. Electronegative sputter ions (e.g. O_2^+) increase the yield of positive ions while electropositive sputter ions (e.g. Cs^+) increase the yield of negative ions. The detection factor η_A is dependent on the setup and the detector. This plethora of factors make quantitative analysis with SIMS complex, requiring standards with the same matrix and known concentrations to obtain sample concentrations from measured intensities. When such standards are not available, a suitable alternative for some secondary ions are isotopes. Since isotopes are the same chemical element, the less accessible parameters in the SIMS equation (η_A, γ) cancel out.

Exchange experiments often make use of the isotopic ratio in order to investigate tracer diffusion. One more abundant isotope already exists in the sample and a second, less abundant isotope is introduced into the sample during an exchange experiment. The less common isotope then functions as a tracer and its secondary-ion profile can be investigated using SIMS. This method is often used (e.g. in this thesis, or refs. 52–54) with oxygen where ^{18}O is used as the tracer for ^{16}O. A large discrepancy in abundance is helpful for differentiating between the isotopes and ^{18}O has a natural abundance of only 0.2 %. It is also available in gas form, which is well suited for exchange experiments. Following the exchange experiments and the SIMS measurements of the exchanged samples, the isotopic ratio $n^*_{^{18}O}$ can be calculated as follows:

$$n^*_{^{18}O} = \frac{I\left(^{18}O\right)}{I\left(^{18}O\right) + I\left(^{16}O\right)}. \tag{3.3}$$

The isotopic ratio $n^*_{^{18}O}$ allows a quantitative interpretation of the data. Since isotopes have a natural abundance, one expects a certain amount of isotope in all samples as a background. Because diffusion experiments often require an additional, external source of isotope n^*_{gas}, the natural background amount of isotope n^*_{bg} is usually subtracted from the isotopic ratio to obtain a corrected n^*_r:

$$n^*_r = \frac{n^*_{^{18}\text{O}} - n^*_{\text{bg}}}{n^*_{\text{gas}} - n^*_{\text{bg}}}.\qquad(3.4)$$

3.2 Others

Other methods of analysis employed in this thesis are listed here without a detailed description as extensive information can be found elsewhere. The depth of SIMS sputter craters and the surface roughness of ceramic samples was measured with interference microscopy.[55] The crystal structure and phase purity of ceramic samples was determined with X-ray diffraction and with grazing incidence X-ray diffraction for thin film samples.[56] Scanning electron microscopy was done to estimate grain sizes for ceramic and thin film samples.[57] For measuring the electrical conductance of ceramic samples, high temperature equilibrium conductance measurements were employed.[58,59]

3.3 Density-Functional-Theory

Quantum-chemical approaches to computational chemistry, also called *ab initio* calculations or first principles calculations, are the premier method of choice for predicting materials' properties.[60] They include all information about the electronic structure of a material in order to achieve great accuracy and reliability. To that end, they aim to solve the Schrödinger equation. In its time-independent form it reads:

$$\hat{H}\Psi = E\Psi. \tag{3.5}$$

$\hat{H}$ is the Hamiltonian, an operator that describes the kinetic energy of all electrons and nuclei and the attractive and repulsive interactions between electrons and nuclei; Ψ is the wave function of a system and the energy E is the eigenvalue of the Hamiltonian. Solving the Schrödinger equation is an extremely complex problem for large systems. To simplify the equation, Born and Oppenheimer[61] introduced an approximation separating the wave functions of the electrons and the nuclei. The nuclei's positions are considered fixed, as they move much slower than the lighter electrons. This changes the Hamiltonian into the electronic Hamiltonian:

$$\begin{aligned}\hat{H} &= \hat{T} + \hat{U} + \hat{V}_{\text{ext}} \\ &= -\sum_{i=1}^{N}\frac{1}{2}\nabla_i^2 + \sum_{i=1}^{N}\sum_{j>i}^{N}\frac{1}{d_{ij}} + \sum_{i=1}^{N}V_{\text{ext}}\left(\mathbf{r_i}\right).\end{aligned} \tag{3.6}$$

$\hat{T}$ is the kinetic contribution of the electrons, $\hat{U}$ is the energy resulting from the interactions between electrons with a distance d_{ij} and $\hat{V}_{\text{ext}}\left(\mathbf{r_i}\right)$ is the interaction energy between a nuclei at position $\mathbf{r}$ and electrons. Analytically solving the Schrödinger equations remains a complex task, even with this simplification. At most, it can be solved for very small systems, such as the hydrogen atom. For larger electronic systems with many electrons N, approximate solutions have to be found.

3.3 Density-Functional-Theory

The Density-Functional-Theory (DFT) is one possible quantum-chemical approach and it is based on the Hohenberg–Kohn theorems.[62] The first of the two theorems states that the electron density ρ unambiguously defines the potential of the nuclei. This means that the energy E of the system is also a function of the electron density:

$$E = E(\rho). \quad (3.7)$$

The second theorem states that the electron density of the ground-state also minimises the energy of said ground-state. Hence, for all states Ψ_t higher than the ground-state Ψ_0 the following holds true:

$$E(\rho_0) = \left\langle \Psi_0 | \hat{H} | \Psi_0 \right\rangle \leq E(\rho_t) = \left\langle \Psi_t | \hat{H} | \Psi_t \right\rangle. \quad (3.8)$$

As a result of these two theorems, DFT can avoid solving the Schrödinger equation for individual electrons and instead solve a Schrödinger-like equation for electron densities. This approach was introduced by Kohn and Sham,[63] who suggested a reference system in which electrons do not interact. These electrons are described by one-electron wave functions, the Kohn–Sham (KS) orbitals ϕ_i. They generate the same electron density as a given system of interacting particles, although they have no physical meaning themselves:

$$\rho(\mathbf{r}) = \sum_{i=1}^{N} |\phi_i(\mathbf{r})|^2. \quad (3.9)$$

The total energy of a system E_ν consisting of these KS orbitals can then be expressed as a functional of the charge density $\rho(\mathbf{r})$:

$$\begin{aligned} E_\nu[\rho(\mathbf{r})] =& T_0[\rho(\mathbf{r})] + E_H[\rho(\mathbf{r})] \\ &+ \int \nu_{\text{ext}}(\mathbf{r})\rho(\mathbf{r})d\mathbf{r} + E_{\text{XC}}[\rho(\mathbf{r})]. \end{aligned} \quad (3.10)$$

Eq. 3.10 consists of four distinct terms, starting with the kinetic energy of the electrons $T_0[\rho(\mathbf{r})]$:

$$T_0[\rho(\mathbf{r})] = \sum_{i=1}^{N} \int d\mathbf{r}\phi_i(\mathbf{r}) \left(-\frac{1}{2}\nabla_i^2\right) \phi_i(\mathbf{r}), \quad (3.11)$$

and the Hartree energy $E_H[\rho(\mathbf{r})]$, describing the Coulomb interaction between electrons:

$$E_H[\rho(\mathbf{r})] = \frac{1}{2} \int \frac{\rho(\mathbf{r})\rho(\mathbf{r}')}{|\mathbf{r}-\mathbf{r}'|} d\mathbf{r} d\mathbf{r}'. \quad (3.12)$$

$\nu_{\text{ext}}(\mathbf{r})$ denotes the external potential acting on the system (generally the Coulomb energy between nuclei and electron) and E_{XC} denotes the exchange-correlation energy among electrons. It combines the non-classical exchange energy between same-spin electrons and the correlation energy of electrons, as well as additional contributions to the kinetic energy.

To then obtain the electron density generated by such an orbital, the KS equations need to be solved until self-consistency is reached. The KS equation for a one-electron orbital is as follows:

$$\left(-\frac{1}{2}\nabla_i^2 + \nu_{\text{eff}}(\mathbf{r}) - \epsilon_i\right) \psi_i(\mathbf{r}) = 0, \quad (3.13)$$

with ϵ_i as the orbital energy of one such KS orbital $\psi_i(\mathbf{r})$ and $\nu_{\text{eff}}(\mathbf{r})$ as the effective KS potential in which the non-interacting electrons move:

$$\nu_{\text{eff}}(\mathbf{r}) = \nu_{\text{ext}}(\mathbf{r}) + \int \frac{\rho(\mathbf{r}')}{|\mathbf{r}-\mathbf{r}'|} d\mathbf{r}' + \nu_{\text{XC}}(\mathbf{r}). \quad (3.14)$$

3.3 Density-Functional-Theory

The exchange–correlation potential $\nu_{XC}(\mathbf{r})$ is the derivative of the exchange–correlation energy E_{XC}:

$$\nu_{XC}(\mathbf{r}) = \frac{\delta E_{XC}[\rho(\mathbf{r})]}{\delta \rho(\mathbf{r})}. \tag{3.15}$$

Exchange–correlation energy and potential are the only unknown parameters in the KS approach, causing the accuracy of the DFT calculations to be dependent on the approach taken to estimate E_{XC}.

One such approach is the Local Density Approximation (LDA),[63] which considers E_{XC} to be solely dependent on local electron densities:

$$E_{XC}^{LDA} = \int e_{XC}[\rho(\mathbf{r})]\rho(\mathbf{r})d\mathbf{r}, \tag{3.16}$$

with $e_{XC}[\rho(\mathbf{r})]$ describing the exchange–correlation energy per particle in a homogeneous electron gas. This term can be separated into an exchange term and a correlation term. The exchange term can be solved analytically, the correlation term can only be approximated through other methods (*e.g.* quantum Monte Carlo methods). A main drawback of the LDA approach is the overbinding problem, referring to a consistent overestimation of binding energies.

As an alternative to the LDA approach, the Generalised Gradient Approximation (GGA) was developed:

$$E_{XC}^{GGA} = \int f_{XC}[\rho(\mathbf{r}), |\nabla \rho(\mathbf{r})|]\rho(\mathbf{r})d\mathbf{r}. \tag{3.17}$$

Here, the functional depends on not only the local density, but also on the gradient $\nabla \rho(\mathbf{r})$ of the density by looking at the surrounding infinitesimal volumes. This means that, contrary to LDA, $e_{XC}[\rho(\mathbf{r})]$ can only be approximated in GGA in the form of $f_{XC}[\rho(\mathbf{r}), |\nabla \rho(\mathbf{r})|]$. This is done through semi-empirical methods, which is why strictly speaking GGA is not an *ab initio* method any longer. Multiple parameterisations exist for f_{XC} (*e.g.* by Perdew, Burke and Ernzerhof (PBE)).[64]

As mentioned in the beginning, large electronic systems are a challenge for DFT due to the computational load involved in solving equations with many parameters. While crystalline solids are always extensive structures with many particles, they are also highly symmetrical and periodic, meaning that the potential is also periodically repeated. This lightens computational requirements a lot, by implementing small unit cells representative of the entire system. Bloch's theorem[65] expresses how the wave function is invariant to the translation vector $\mathbf{T}$:

$$\Psi(\mathbf{k}, \mathbf{r} + \mathbf{T}) = \exp(i\mathbf{k}\mathbf{T})\Psi(\mathbf{k}, \mathbf{r}). \qquad (3.18)$$

The wave vector is described as a point in reciprocal space $\mathbf{k}$. A set of these points is sampled to obtain an accurate picture of the electron density. This set can be generated with the Monkhorst–Pack scheme[66] or a different method, but the Monkhorst–Pack scheme guarantees equally spaced points.

To describe the wave functions in solids, generally plane-wave basis sets are used, as the periodic structure is well suited to the expansion of the wave function with the plane waves $\exp(i\mathbf{k}\mathbf{T})$. Their accuracy can be controlled by choosing a suitable cut-off energy. The downside to the plane-wave basis set is the description of regions near nuclei, which cause strong oscillations in the wave function and thus need a large number of plane waves to be described. Augmented plane-wave methods use approximations such as the muffin-tin approximation to describe these oscillations with atomic wave functions, but need a lot of computational resources. Alternatively, the oscillations can be removed by using pseudo potentials. Pseudo potentials describe the total charge and potential of the nucleus and the core electrons correctly, but do not describe the electrons explicitly, resulting in much easier calculations and no oscillations at the price of losing information about the core electrons. Projector Augmented Wave (PAW)[67] methods generate such pseudo potentials during the calculation, adapting to the electronic environment and resulting in better accuracy than normal augmented plane-wave methods.

If one wants to use DFT to calculate the migration enthalpy ΔE_{mig} of a specific ion jump, Nudged Elastic Band (NEB) calculations[68] can be useful. Generally all calculations are static, *i.e.* at 0 K. Calculating the total energy of an intermediary ion configuration during the jump to obtain an energy barrier is not possible, because the ion will relax back into its equilibrium lattice positions. NEB circumvents this by connecting the intermediate configurations or images with the initial and final configurations through elastic bands. These bands apply a force on neighbouring configurations, which prevents them from relaxing into the initial/final configuration. The spring forces act only along the migration path and all forces perpendicular to that path are set to zero. Ideally, the maximum-energy image corresponds to the saddle-point configuration, which defines the migration barrier. However, this is not guaranteed for NEB, as the images are spaced equally. Thus, Climbing-Image NEB (CI-NEB) was developed by Henkelman *et al.*,[69,70] an expanded method which drives the highest energy image up to the saddle-point. As a consequence, the images are not spaced equally any longer but the calculations find a saddle-point configuration as long as they converge.

3.4 Molecular Dynamics

While quantum-chemical approaches, such as DFT, provide great accuracy and reliabe predictions for material parameters, they also require significant computational resources, leading to an upper limit of system size. Once finite temperatures become a point of interest, the computational demand increases again. However, simulations containing thousands of ions over a time span of several nanoseconds and at significant finite temperatures are often required for the investigation of ion diffusion.[71] These requirements are beyond the current capabilities of DFT-based methods. Atomistic approaches, however, are capable of fulfilling these requirements by avoiding an exact description of the electronic structure of chemical systems in order to lower computational demands. Instead, ion interactions are parameterised through so-called Empirical Pair Potentials (EPP). Classical Molecular Dynamics (MD) are based on Newton's second law, where the force affecting a particle in a system is proportional to the positional derivative of the particles potential energy U_i:

$$-\frac{\mathrm{d}U_i}{\mathrm{d}r_i} = m_i \frac{\mathrm{d}^2 r_i}{\mathrm{d}t^2}, \tag{3.19}$$

with r_i as the position vector of the referred particle and m_i its mass. The second derivative of the position after time $\frac{d^2r}{dt^2}$ is the acceleration a in Newton's laws of motion. When position and influencing force of a particle are known for a specific point in time t, then position and force can be determined for a different point in time $t \pm \Delta t$, as long as the time step Δt is very small.

$$r_{i\pm 1} = r_i \pm \frac{\partial r}{\partial t}(\Delta t) + \frac{1}{2}\frac{\partial^2 r}{\partial t^2}(\Delta t)^2 \pm \frac{1}{6}\frac{\partial^3 r}{\partial t^3}(\Delta t)^3 + \ldots \tag{3.20}$$

The dependence on the size of Δt is explained with Eq. 3.20, which shows the Taylor series for $r_{i\pm 1}$. The derivatives of the position after time shown are, in order, velocity v, acceleration a and change in acceleration $\frac{da}{dt} = b$. To predict a future position of a particle, the actual and previous position need to be known:

$$r_{i+1} = (2r_i - r_{i-1}) + a_i(\Delta t)^2 + \ldots \tag{3.21}$$

Since Δt is small, only the linear and the square term of the Taylor series are relevant. This is known as the Verlet algorithm.[72] The Verlet algorithm is often expanded upon by including the velocity of the particles, resulting in the velocity Verlet algorithm.[73] Eq. 3.21 then reads as follows:

$$r_{i+1} = (2r_i - r_{i-1}) + a_i(\Delta t)^2 + v_i \Delta t \tag{3.22}$$

In this algorithm the velocity v_{i+1} is calculated by taking the actual velocity v_i, half of the actual acceleration a_i and half of the future acceleration a_{i+1} into account:

$$v_{i+1} = v_i + \frac{1}{2}a_i \Delta t + \frac{1}{2}a_{i+1} \Delta t \tag{3.23}$$

The relation of acceleration a_i and potential energy U_i is described in Eq. 3.19. At the beginning $t = 0$ of a MD simulation, the position and acceleration of all particles are unknown, so an approximation based on Δt is done:

$$r_{-1} = r_0 - \frac{\partial r_0}{\partial t}\Delta t. \tag{3.24}$$

3.4 Molecular Dynamics

To this end, all particles are assigned a starting velocity v_0 before the start of a simulation, which is randomised appropriate to a specified temperature T. The sum of all particles and the thermodynamic system quantities in a simulation is the ensemble. In MD, the total energy $U(r)$ of a system is equal to the sum of all interaction energies U_{pair} between two particles (in the system), with N as the total number of particles in the system.

$$U(r_1, r_2, \ldots, r_n) = \sum_{i=1}^{N-1} \sum_{j>i}^{N} U_{\text{pair}}(r_i, r_j) \qquad (3.25)$$

The basic ensemble is called the microcanonical ensemble (NVE). Particle number N, volume V and total energy E are kept constant in this ensemble while temperature and pressure may vary during simulation. This is not suitable for many different types of simulations, so different ensembles exist. There is the canonical ensemble (NVT) where particle number N, volume V and temperature T are kept constant while pressure and energy may vary. Lastly, there is the isothermal–isobaric ensemble (NpT) where particle number N, pressure p and temperature T are kept constant while volume and energy may vary. Temperature and pressure in the NVT or NpT ensembles are varied through the use of a Nosé–Hoover thermo- or barostat, which shall not be explained here in detail.[74,75] The kinetic energy E_{kin} of the entire system is dependent on the temperature T of the system:

$$\langle E_{\text{kin}} \rangle = \frac{1}{2}(3N_{\text{atoms}})k_{\text{B}}T. \qquad (3.26)$$

The potential energy $U(r)$ of the system, described in Eq. 3.25, and the sum of kinetic energy over all particles, form the total energy of the system:

$$E_{\text{tot}} = \sum_{i=1}^{N} \frac{1}{2} m_i v_i^2 + U(r). \qquad (3.27)$$

The interactions between particles are described by EPP. These contain potential functions which are usually comprised of a term for long-range electrostatic interactions and a short-range term. For ionic systems, the long-range term generally describes Coulomb interactions, where the electrostatic potential energy $U_{\text{pair, lr}}(r_i, r_j)$ between two particles is as follows:

$$U_{\text{pair, lr}}(r_i, r_j) = \frac{1}{4\pi\epsilon_0}\left(\sum_{i=1}^{N}\sum_{j>i}^{N}\frac{q_i q_j}{r_{ij}}\right), \tag{3.28}$$

with q denoting the charge of a particle and r_{ij} the distance between two particles. ϵ_0 refers to the vacuum permittivity. The short-range terms are empirically fitted to known data from experiments and *ab-initio* calculations, since they are usually very ion-specific. That also means that they are not always suited to a specific problem, if they are not fitted with this problem in mind. Buckingham potentials are one of the most common formalisms of these terms:

$$U_{\text{pair, sr}}(r_i, r_j) = A e^{-B r_{ij}} - \frac{C}{r^6}, \tag{3.29}$$

where A, B and C are empirical parameters different for each potential. For HfO_2 in this thesis, potentials derived by Lewis and Catlow[76] are used, as they have shown to be well suited to reproduce ion migration in various ionic oxides.[25,29,77–80]

MD is very well suited to investigate diffusion processes, since almost all parameters needed are either given by the simulation input or obtained through the simulation. The mean squared displacement $\langle \mathbf{r}^2 \rangle$ over all ions of a single species is directly proportional to the tracer diffusion coefficient D^*:

$$\langle \mathbf{r}^2 \rangle = 6 D^* t. \tag{2.19 revisited}$$

The tracer diffusion coefficient D^* calculated this way can then be used to obtain the enthalpy of migration $\Delta H_{\text{mig,ion}}$ for any species in the simulation:

$$\ln D^* = \ln D^0 - \frac{1}{T}\frac{\Delta H_{\text{mig}}}{k_B}. \tag{2.28 revisited}$$

Additionally, an external electrical field E can be applied during an MD simula-

3.4 Molecular Dynamics

tion in the form of a simple external force $F = qE$. This force is applied to each atom with charge q according to the specifications of the user.

To conclude, MD is a very useful tool for simulating ion migration in solids and often gives results comparable with experimental results. Please refer to chapter 2.4 for a more extensive description.

Chapter 4

Oxygen Diffusion in HfO$_2$

Parts of this chapter have previously been published:

M. P. Mueller and R. A. De Souza, "SIMS Study of Oxygen Diffusion in Monoclinic HfO$_2$" *Appl. Phys. Lett.* **2018**, 112, 51908.[81]

4.1 Introduction

Due to the challenges involved in fabricating dense ceramic samples of monoclinic HfO$_2$ (m-HfO$_2$), a direct investigation of the anion diffusion, for which dense samples are a necessity, has not been done before. These challenges stem from the close proximity of the needed sintering temperatures to the temperatures of phase transition in HfO$_2$. Previously, experimental evidence was limited to investigations of the space-charge limited current[82] or to atomistic simulations, where it is unclear how close those results are to reality. Diffusion experiments on the other hand directly follow the motion of one ion species and thus depict reality directly. This chapter thus describes the investigation of oxygen diffusion in dense ceramics of m-HfO$_2$ using ^{18}O/^{16}O isotope exchange and subsequent Time-of-Flight Secondary Ion Mass Spectrometry (ToF-SIMS) analysis. The obtained isotope profiles were then analysed using Finite-Element-Method (FEM) simulations.

4.2 Experimental Details

Ceramic samples of HfO_2 were prepared by sintering commercial powder (99.95 % purity excluding zirconium, Alfa Aesar®) at $T = 1823\,\text{K}$ for 72 h. In preparation for the pressing process, the commercial powder was ball milled in an ethanol-filled Teflon cup with 5 mm ZrO_2 balls for 2 h in a planetary ball mill of type Pulverisette 7 (Fritsch GmbH, Idar-Oberstein, Germany). Following the ball milling, the ethanol–HfO_2 suspension was evaporated and the remaining HfO_2-powder dried. 0.65 g of the dried powder were then ground in an agate mortar and pressed in an uniaxial press. The applied force was 20 kN for 10 min on a round pellet of 10 mm diameter. The fabricated pellets were then sealed airtight and pressed once again, this time in an isostatic press at 200 kN for 45 min. Afterwards all pellets were sintered in a $MoSi_2$ furnace in air inside a closed Al_2O_3 vessel. For the sintering, the samples were heated up to 1373 K with a heating ramp of 90 K/h, then immediately up to 1828 K at a ramp of 30 K/h. The temperature was then held for 72 h and cooled down with the same ramps. Several pellets were prepared and only the pellets with the highest density were used for isotope exchange experiments.

The density of the used pellets was at least 96 % of the theoretical density, measured by Archimedes' principle, and thus sufficient for isotope exchange experiments (at least 95 % are required).[83] X-ray diffractograms obtained with a Theta–Theta diffractometer (STOE & Cie GmbH, Darmstadt, Germany) indicated single-phase m-HfO_2. Fig. 4.1 shows a sample diffractogram for one of the samples. Evidently a single monoclinic phase exists.

The grain-size was estimated from Scanning Electron Microscopy (SEM) images taken with a ZEISS (LEO) 1450VP system (Carl Zeiss AG, Oberkochen, Germany) to be $d_{gr} \approx 3\,\mu\text{m}$. Since HfO_2 is an isolator, a thin layer of gold was sputtered onto the samples, to improve charge distribution and SEM resolution. Fig. 4.2 shows an exemplary SEM scan of an unpolished sample.

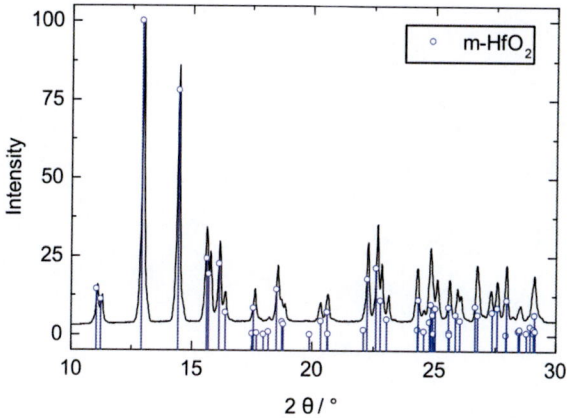

Figure 4.1: Diffractogram of a sintered ceramic m-HfO$_2$ sample. The black line represents the measured data and blue lines indicate the position of reflections caused by m-HfO$_2$.[84]

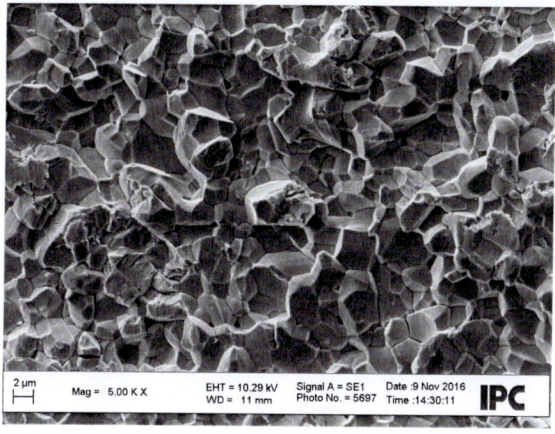

Figure 4.2: SEM image of an unpolished, sintered ceramic m-HfO$_2$ sample.

Prior to the diffusion experiments, the samples were polished with successive grades of diamond suspension to a surface roughness of $R_q = \pm 15\,\text{nm}$ (over an area of $480\,\mu\text{m} \times 736\,\mu\text{m}$), as indicated by interference microscopy (NT1100, Veeco Instruments Inc., New York, USA); for the areas relevant to ToF-SIMS analysis

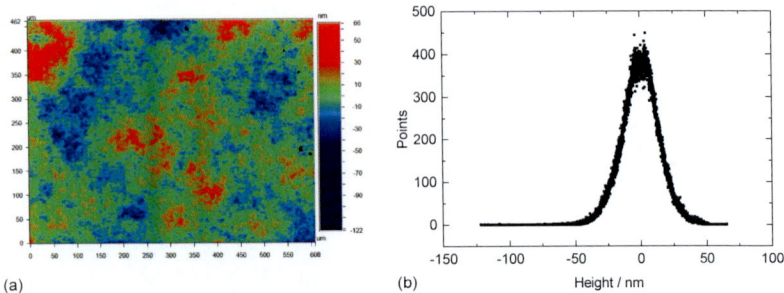

Figure 4.3: (a) Interferogram of a sintered and polished ceramic m-HfO$_2$ sample; (b) Histogram of sintered and polished ceramic m-HfO$_2$ sample.

(100 μm × 100 μm), the roughness is $R_q = \pm 10$ nm. Fig. 4.3 shows an interferogram and the respective histogram of a sample surface.

According to the suppliers, the main impurities present in the HfO$_2$ powder are: Zr (2000 ppm), Ca (220 ppm), Si (30 ppm), Al (25 ppm), Fe (25 ppm) and Ni (18 ppm). This data is used to estimate the effective concentration of oxygen vacancies in the HfO$_2$ samples. Assuming, first, that the impurities all substitute for Hf; that Zr and Si are isovalent to Hf ($Zr_{Hf}^\times$, $Si_{Hf}^\times$); that the other impurities are present as Ca_{Hf}'', Al_{Hf}', Fe_{Hf}' and Ni_{Hf}''; an effective acceptor concentration of $c_{Ac} \approx 1.5 \times 10^{19}$ cm^{-3} is obtained. With the reasonable electroneutrality condition, $c_{Ac} = 2c_v$, as another assumption, an oxygen-vacancy concentration of $c_v \approx 7.5 \times 10^{18}$ cm^{-3} is then estimated. It is surmised that this value represents a maximum possible concentration; the actual concentration may be lower, and this is discussed later. The concentration of oxygen interstitials can be safely neglected[85] on account of m-HfO$_2$ being acceptor-doped and the anti-Frenkel formation enthalpy being so high ($\Delta H_{aF} = 4.7 - 8.0$ eV):[86,87] $c_i \sim c_v^{-1} \exp\left(-\Delta H_{aF}/k_B T\right)$.

In a tracer diffusion experiment,[83,88,89] a sintered HfO$_2$ sample was first annealed in molecular oxygen of normal isotopic abundance at the given temperature of interest (and at an oxygen partial pressure of $pO_2 = 200$ mbar) for ten times the duration of the exchange anneal for two reasons: (i) to bring the sample into chemical equilibrium with gaseous oxygen (to ensure that an oxygen exchange experiment occurs, *i.e.* the introduction of ^{18}O and the concurrent removal of an equivalent amount of ^{16}O) and (ii) to yield a relatively stable surface (to ensure

4.2 Experimental Details

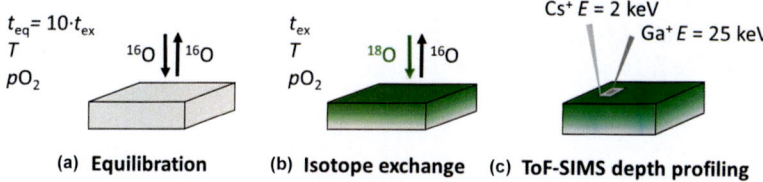

Figure 4.4: Sketch of an oxygen isotope exchange experiment. (a) Equilibration process of samples with long equilibration time t_{eq}; (b) Actual isotope exchange with shorter exchange time t_{ex}. ^{18}O is introduced to the sample and ^{16}O is removed; (c) ToF-SIMS depth profiling with primary ion type and energy and sputter ion type and energy.

that any changes in sample composition will occur predominantly during the pre-anneal). This equilibration step is displayed in Fig. 4.4(a). The sample was then subjected to an exchange experiment in a large volume of 98 % ^{18}O-enriched molecular oxygen at a temperature in the range $573 \leq T/\text{K} \leq 973$ for a time $20 \leq t/\text{min} \leq 90$ (as seen in Fig. 4.4(b)), and then quenched to room temperature.

Fig. 4.4(c) represents the measurement step of the isotope exchange experiment, the SIMS measurement. More detailed information about this analysis method are contained in chapter 3.1. ToF-SIMS depth profiles were obtained [52,90] on a TOF.SIMS IV machine (ION-TOF GmbH, Münster, Germany). 25 keV Ga$^+$ ions, rastered over an area of 100 µm × 100 µm, were used as primary ions to generate secondary ions. 2 keV Cs$^+$ ions were used for sputter etching the sample surface, typically over an area of 300 µm × 300 µm. Charge compensation was accomplished with a beam of low-energy (<20 eV) electrons. Measurements were performed with a ToF cycle time of 60 µs in either bunched mode or burst mode (with 7 bursts),[52] and negative secondary ions were detected. The isotope fraction is calculated from the secondary-ion intensities according to $n^*(x) = I\,(^{18}\text{O}^-)/(I\,(^{18}\text{O}^-) + I\,(^{16}\text{O}^-))$. Crater depths were determined post-analysis by interference microscopy.

The interpretation of the isotope depth profiles happens by solution of the diffusion equation. For the initial and boundary conditions defined by the experimental setup, a solution of the diffusion equation is already known:

$$n_r^*(x) = \mathrm{erfc}\left(\frac{x}{2\sqrt{D^*t}}\right) - \exp\left(\left(\frac{k^*}{D^*}\right)x + \left(\frac{k^*}{D^*}\right)^2 D^*t\right)$$
$$\mathrm{erfc}\left(\frac{x}{2\sqrt{D^*t}} + \frac{k^*}{D^*}\sqrt{D^*t}\right), \qquad \text{(2.18 revisited)}$$

where D^* is the oxygen tracer diffusion coefficient and k^* is the oxygen surface exchange coefficient. $n_r^*(x) = (n^*(x) - n_{bg}^*)/(n_{gas}^* - n_{bg}^*)$ is the corrected oxygen isotope fraction with n_{gas}^* as the isotope fraction of ^{18}O in the annealing gas and n_{bg}^* as the background isotope fraction in the sample.

4.3 Results

All ^{18}O diffusion profiles obtained for m-HfO$_2$ ceramics by ToF-SIMS depth profiling showed two features. One such profile is plotted in Fig. 4.5. The first feature (situated closer to the surface and termed F1) extended in this case approximately 100 nm from the surface and cannot be described well by Eq. (2.18) (see inset of Fig. 4.5). The second feature (situated further from the surface and termed F2) extended in this case up to 500 nm and can, in contrast, be described well by Eq. (2.18). For other isotope profiles (not shown), F1 varied in extent between 80 nm and 500 nm but in no apparent systematic manner with annealing time or temperature, whereas F2, extending up to 1500 nm, was longer for isotope anneals at higher temperatures or for longer times. It is possible that F1 is a measurement artefact, resulting from some combination of ion beam mixing, charging effects, non-steady-state sputtering and surface roughness, but all these possibilities can

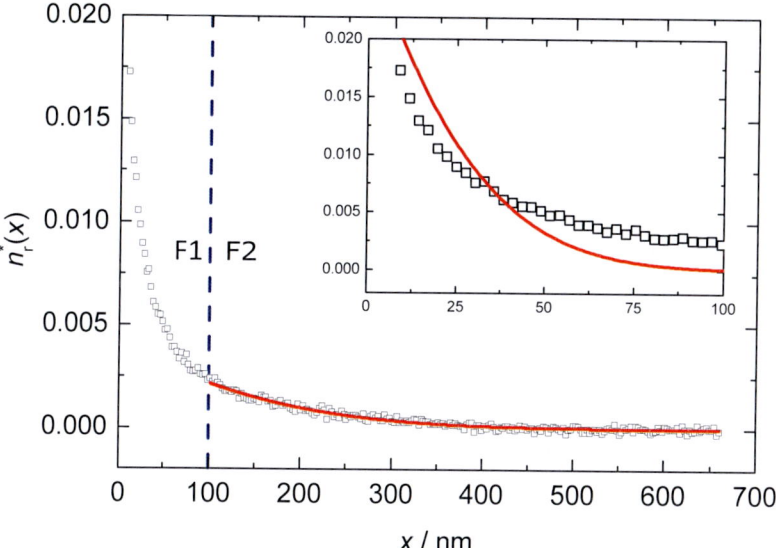

Figure 4.5: Corrected ^{18}O isotope fraction versus depth obtained by ToF-SIMS analysis of an m-HfO$_2$ ceramic after an exchange anneal at $T = 623$ K and $pO_2 = 200$ mbar for 90 min. Squares: experimental data. Line: Fitted curve, Eq. 2.18. Inset shows a fit to data over the first 100 nm.

be discounted either because F1 is too long (up to 500 nm) or because both (chemically identical) isotopes would be affected in the same way and thus the isotope ratio n_r^* would be unaffected. Therefore, it is concluded that these profiles with two features are characteristic of the samples themselves.

4.4 Discussion

The traditional explanation for diffusion profiles in polycrystals with two features is Harrison type B diffusion kinetics:[46,90] the first feature is due to bulk diffusion; the second feature is due to combined fast diffusion along grain boundaries and slow bulk diffusion out of the boundaries. For Harrison type B kinetics to be valid, the diffusion length should lie between the width of the grain boundary, d_{gb}, and the grain size d_{gr}. This condition is indeed fulfilled for all profiles obtained, with F1 ($\sim 10^2$ nm) being much larger than the assumed grain-boundary width ($d_{gb} \approx 1$ nm) and much smaller than the measured grain size ($d_{gr} \approx 3$ μm). While this observation is in favour of Harrison-type-B diffusion kinetics, it is only a check of validity. The fact that F1 cannot be described by Eq. (2.18) in any profile but F2 can, and the fact that F1 does not appear to vary with annealing time or temperature but F2 does, both count strongly against this explanation.[91,92]

An alternative, and more recent, explanation for a diffusion profile in an acceptor-doped oxide with two features is slow diffusion through an equilibrium space-charge layer (in which the oxygen vacancies responsible for diffusion are depleted) followed by faster diffusion in an homogeneous bulk phase.[49,52,93,94] The diffusion equation is solved numerically in two dimensions with FEM simulations in order to obtain a description of the experimental data. These simulations were implemented in COMSOL Multiphysics® (COMSOL AB, Stockholm, Sweden). Fitting parameters in addition to D^* and k^*, in this case, are the space-charge potential Φ_0, the acceptor concentration c_{Ac} and the relative dielectric permittivity κ. While this model delivers reasonably good descriptions of the experimental profiles—indicating that oxygen diffusion in F1 is much slower than in F2—, there are two problems with it. First, the fitting parameters required to describe the profiles varied strongly from sample to sample. Second, good descriptions of feature F1 were only possible with physically unreasonable values of $c_{Ac} \sim (10^{14} - 10^{16})$ cm^{-3} (*i.e.* three to five orders of magnitude lower than the

4.4 Discussion

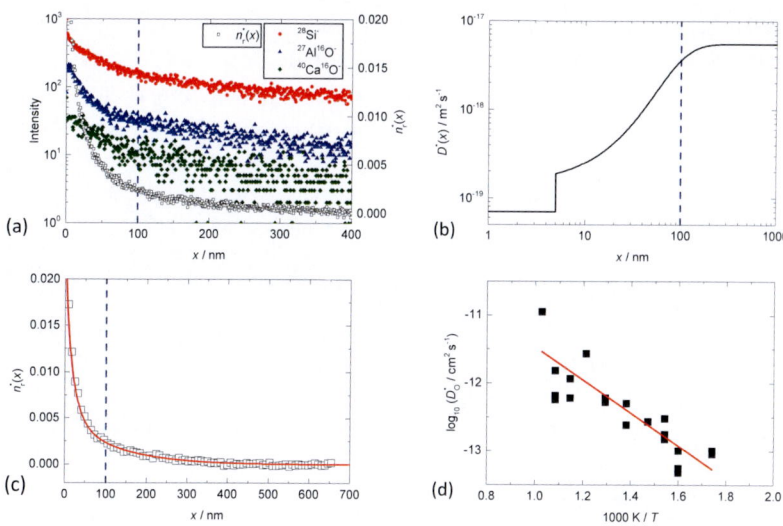

Figure 4.6: (a) $n_r^*(x)$ (squares) compared with SI intensity profiles of ^{28}Si$^-$ (circles), ^{27}Al^{16}O$^-$ (triangles) and ^{40}Ca^{16}O$^-$ (diamonds) for an m-HfO$_2$ ceramic after an ^{18}O/^{16}O exchange anneal at $pO_2 = 200$ mbar and $T = 623$ K for 90 min. (b) Assumed variation in tracer diffusion coefficient versus depth. (c) Experimental isotope profile (squares) compared with a simulated curve (line) calculated from $D^*(x)$ shown in (b). (d) Oxygen tracer diffusion coefficients for m-HfO$_2$ as a function of inverse temperature.

initial estimate). Together, these two problems argue strongly against an equilibrium space-charge layer.

Hafnia's structural and chemical twin, zirconia (ZrO$_2$), is known to be plagued by segregation of silicium to sample surfaces.[95–99] Furthermore, two different studies found that phase separation between HfO$_2$ and SiO$_2$ occurs, in which SiO$_2$ migrates to the surface of HfO$_2$.[100,101] Goncharova et al.[102] suggested that SiO$_2$ segregation to the surface suppresses oxygen exchange in thin HfO$_2$ films. Therefore, the ToF-SIMS raw data for impurity species in the m-HfO$_2$ ceramics was examined. Sufficient secondary-ion intensity (i.e. well above background level) was observed for the ^{28}Si$^-$, ^{27}Al^{16}O$^-$ and ^{40}Ca^{16}O$^-$ species. These impurity species were also found when positive secondary ions were detected, but with less intensity. Fig. 4.6 (a) compares the normalised intensity profiles of these species with the oxygen isotope profile, $n_r^*(x)$, for one sample. Similar behaviour for the impurity and isotope profiles is perceived, particularly with regard to the extent of feature F1 (the region to the left of the dotted line). Note that the forms of the impurity and isotope profiles will differ because the former refers to the composition, while the latter reflects the local oxygen diffusion coefficient. This behaviour is seen in all m-HfO$_2$ samples analysed. It is therefore concluded, that these substantial compositional changes are responsible for the slower diffusion of oxygen in F1, with F2 corresponding to faster diffusion in m-HfO$_2$.

On a purely empirical basis, it was found that the best descriptions of the isotope profiles required three different regions: a region of constant low diffusivity and variable extent; an intermediate regime of variable diffusivity and variable extent; and a semi-infinite region of high diffusivity (see Fig. 4.6(b). N. b.: logarithmic abscissa scale). That is, the first two regions together comprise F1. The extent of the first region was found to vary between 5 and 25 nm in the descriptions, with higher temperatures generally being characterised by a thicker region. The very low rate of oxygen diffusion in this region is attributed to the presence of a silicate phase at the very surface.[102–104] The intermediate region is ascribed to substantial changes in the composition of HfO$_2$ and possibly to the silicate phase being of variable thickness. Although the level of impurities detailed by the supplier indicates only a small amount of Si (30 ppm), additional silicon may have been introduced accidentally during the polishing step of sample preparation. Nevertheless, even 30 ppm of silicon is sufficient to create a SiO$_2$ layer approximately 40 nm thick, with a sample mass of 0.63 g, a sample area of 0.51 cm^2 and a sample height of 0.13 cm. This calculation is assuming the best case, but even thicker layers are conceivable if a silicate phase forms, containing hafnia, alumina and calcia. In any case, this thickness is of the same order of magnitude as the first region, of constant low diffusivity, described in the profiles. The third region is considered to refer to oxygen diffusion in m-HfO$_2$. It needs to be emphasised that, although the sole interest of this chapter is in oxygen diffusion in m-HfO$_2$, the correct extraction of D^* requires a theoretical description of the entire isotope profile, since the processes occur in series. Fig. 4.6(c) compares the isotope profile predicted with $D^*(x)$ from Fig. 4.6(b) with the experimental data. Repeating this procedure for all measured isotope profiles yielded the diffusion data plotted in Fig. 4.6(d) as a function of inverse temperature. From this plot an activation enthalpy of oxygen tracer diffusion in m-HfO$_2$ of $\Delta H_{D^*} = (0.48 \pm 0.12)$ eV (where the error corresponds to twice the standard deviation) is obtained.

4.4 Discussion

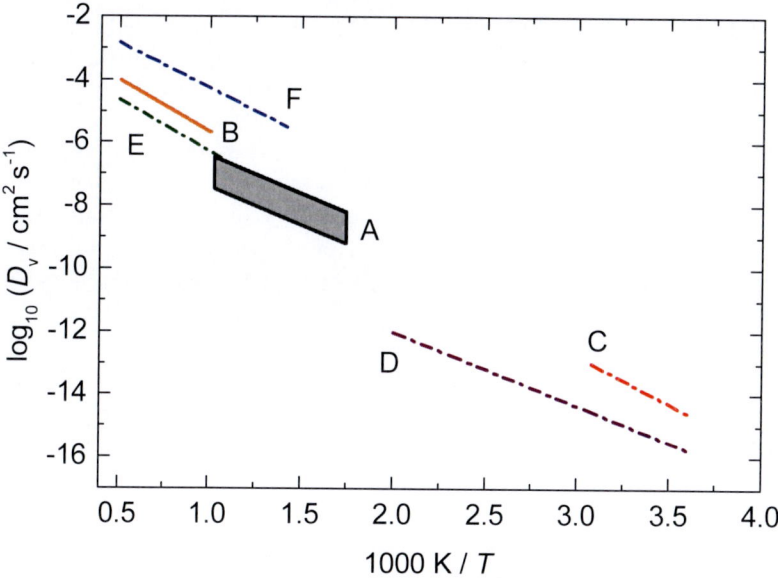

Figure 4.7: Comparison of oxygen-vacancy diffusion coefficients: (A) m-HfO$_2$, this work; (B) m-HfO$_2$, MD simulations;[29] (C,D) a-HfO$_x$, current–voltage studies;[82] (E,F) a-HfO$_x$, DFT-MD simulations.[17]

The comparison of diffusion data is best conducted at the level of defect diffusivities, since for low defect concentrations, the defect diffusivity is independent of defect concentration. Specifically in the present case, the measured tracer diffusion coefficient of oxygen (D^*) is related to the diffusion coefficient of oxygen vacancies (D_v) through $D^* = f^* D_v c_v / N$ (compare with Eq. 2.11), where f^* is the tracer correlation coefficient and N is the number of anion sites. Since f^* is not known for vacancy migration on the anion sublattice of m-HfO$_2$, as a first approximation the value of $f^* = 0.65$ for anion migration in a cubic fluorite-structured oxide is used.[23] A maximum value of $c_v^{max} \approx 7.5 \times 10^{18}$ cm^{-3} was estimated above from the level of impurities detailed by the suppliers. This level represents a maximum value because the two main acceptor-dopant species, Al and Ca, are highly enriched within F1, meaning that their concentrations within F2 are diminished. It is not possible to say, however, from the SIMS data (due to the SIMS matrix effect) exactly how large these variations are. As a conservative estimate, a lower limit of impurities is put in place, and thus of oxygen vacancies, one order of magnitude lower, $c_v^{\min} \approx 7.5 \times 10^{17}\,\text{cm}^{-3}$. In this way, the region denoted A in Fig. 4.7 is obtained. The activation enthalpy of vacancy migration is $\Delta H_{D_v} = (0.48 \pm 0.12)\,\text{eV}$ (since c_v is fixed by c_{Ac} through electroneutrality).

Oxygen diffusion coefficients for HfO_2 are quite under-reported. The few available resources either obtained values indirectly by modelling the current–voltage behaviour of thin films[82] or predicted values by Molecular Dynamics (MD) simulations.[17,29,105] The only other set of data for vacancy diffusion in m-HfO_2 comes from MD simulations with empirical pair-potentials reported by Schie et al.[29] (B), the other simulations refer to amorphous HfO_2. As seen in Fig. 4.7, there is relatively good agreement both in terms of the absolute magnitude and in terms of the activation enthalpy of vacancy migration, with Schie et al.[29] obtaining $\Delta H_{D_v} = 0.66\,\text{eV}$. Density-Functional-Theory (DFT) calculations[106,107] yield activation enthalpies of $0.69\,\text{eV}$ and $0.71\,\text{eV}$ for vacancy migration in m-HfO_2.

The data obtained from other studies[17,82] shown in Fig. 4.7 does not differ by many orders of magnitude but also refers to oxygen diffusion in amorphous HfO_x (a-HfO_x) systems, rather than m-HfO_2. The lack of large differences between the data for amorphous and monoclinic phases may be ascribed to the migration barriers being relatively similar in the two systems,[29] on account of the short-range-order of the ions being similar, too.

4.5 Summary

The diffusion of oxygen in dense ceramics of m-HfO_2 was studied by means of ($^{18}O/^{16}O$) isotope exchange annealing and subsequent determination of the isotope depth profiles by SIMS. All measured isotope profiles showed complicated behaviour in exhibiting two features: the first feature, closer to the surface, was attributed mainly to slow oxygen diffusion in an impurity silicate phase; the second feature, deeper in the sample, was attributed to oxygen diffusion in bulk m-HfO_2. The activation enthalpy of oxygen tracer diffusion in bulk m-HfO_2 was found to be $\Delta H_{D^*} \approx 0.5\,\text{eV}$.

Chapter 5

Cation Diffusion in HfO$_2$

Parts of this chapter have been submitted for publication:

M. P. Mueller, K. Pingen, A. Hardtdegen, S. Aussen, A. Kindsmueller, S. Hoffmann-Eifert, and R. A. De Souza, "Cation Diffusion in Polycrystalline Thin Films of Monoclinic HfO$_2$ deposited by Atomic Layer Deposition" *APL Mater.* **2020**, 8, 081104. [108]

5.1 Introduction

To reiterate on the introduction of this thesis, relative diffusion rates in oxides present an important issue in examining the diffusion of less mobile ions. Looking at all the AO_2 oxides that adopt the fluorite and related structures, such as HfO$_2$, ZrO$_2$, CeO$_2$, ThO$_2$, UO$_2$ and PuO$_2$, it becomes apparent that they all have mobile oxide ions and relatively immobile cations.[26,109–111] Indeed, solid solutions based on ZrO$_2$ or CeO$_2$ are utilised as oxide-ion conducting electrolytes.[112–115]

This chapter looks at cation diffusion in HfO$_2$ and these cations are immobile compared to the mobile oxide ions. Studying experimentally the diffusion of a crystal's 'immobile' ions is difficult because the number of defects responsible for diffusion is generally low and the activation enthalpy of defect migration is generally high, which means that diffusion coefficients are necessarily very small. High temperatures and long diffusion times are thus needed to achieve measurable diffusion lengths, conditions that are often accompanied by complicating effects (*e.g.* diffusion specimens showing morphological changes).

If one manages to obtain diffusion data of sufficient quality, one faces the further problem of interpreting the measured activation enthalpy of diffusion, ΔH_D. Even in the simplest case, ΔH_D is the sum of two individual enthalpies, ΔH_{mig}, the activation enthalpy of defect migration, and ΔH_{gen}, the enthalpy of defect generation:

$$\Delta H_D = \Delta H_{\text{mig}} + \Delta H_{\text{gen}}. \tag{5.1}$$

ΔH_{gen} describes how the concentration of the defect responsible for diffusion, c_{def}, changes with temperature T:

$$\Delta H_{\text{gen}} = -k_{\text{B}} \left[\frac{\text{d} \ln c_{\text{def}}}{\text{d}(1/T)} \right]. \tag{5.2}$$

For minority defects neither ΔH_{mig} nor ΔH_{gen} is independently accessible from experiment.

Simulation techniques, based either on Density-Functional-Theory (DFT) calculations or on calculations with Empirical Pair Potentials (EPP), can prove helpful by providing both ΔH_{mig} and the relevant defect energies entering ΔH_{gen}. Help is not always guaranteed, however. In the case of cation diffusion in Gd-doped CeO_2, for example, the experimental value of $\Delta H_D \approx 5.5\,\text{eV}$ is substantially lower than the sum of the (DFT) computational values for vacancy migration, with $\Delta H_{\text{mig,v}} = 4.4\,\text{eV}$ and $\Delta H_{\text{gen,v}} = 6.9\,\text{eV}$.[26,38,116] Similarly for UO_2, the experimental value of $\Delta H_D \approx 5.6\,\text{eV}$ is substantially lower than the (EPP) computational values of $\Delta H_{\text{mig,v}} = 6\,\text{eV}$ and $\Delta H_{\text{gen,v}} = 6.5\,\text{eV}$.[117] Values for cation interstitials show even larger discrepancies.

The interpretation of experimental values of ΔH_D would be simplified if the concentration of the defects responsible for diffusion could be fixed over an appropriate range of temperatures, so that ΔH_{gen} goes to zero. For majority defects, a suitable dopant and suitable thermodynamic conditions may serve to fix c_{def} (*e.g.*, in Gd-doped CeO_2 under oxidising conditions the concentration of oxygen vacancies is fixed by the gadolinium dopant concentration) and make investigation easier, but that is not the case for minority defects. For minority defects, generally

5.1 Introduction

connected with the diffusion of 'immobile' ions, a variety of intrinsic and extrinsic defect reactions determines c_{def}. That is, the concentrations of minority defects are dictated by thermodynamics and subsequently make the investigation of diffusion of 'immobile' ions even more challenging. In general, c_{def} will therefore vary with temperature, giving rise to non-zero ΔH_{gen}. For the case of Gd-doped CeO_2 mentioned above, with gadolinium fixing the oxygen vacancy concentration, ΔH_{gen} for cation vacancies is equal to ΔH_{Sch} (the enthalpy of Schottky disorder). One way to avoid this problem, then, is to study samples that are not in equilibrium, so that thermodynamics is removed from the problem. This can be achieved by investigating cation diffusion in metastable samples, in which the concentration of the minority defects is fixed through kinetics, i.e., through the preparation procedure.

Such samples can be fabricated by Atomic Layer Deposition (ALD), the established industrial process for growing nanometer-thin, homogeneous insulating HfO_2 thin films. The primary advantage for this thesis is that this method offers the possibility of producing at low temperatures polycrystalline films of monoclinic HfO_2 (m-HfO_2).[118] The low temperatures provide the sluggish kinetics required so that equilibrium is not achieved for the cation sublattice. An additional advantage is that ALD produces films that are free of Si (whose presence complicates the study of diffusion in AO_2 oxides.[81,95,102,119]) The use of HfO_2 also offers the advantage of using zirconium as a chemically similar species to study cation diffusion rather than using expensive (and hard to process) hafnium isotopes. Thus, zirconium diffusion in HfO_2 was experimentally studied in thin films of monoclinic HfO_2 (m-HfO_2) produced by ALD in this chapter. Diffusion annealing experiments were done using a ZrO_2 layer on top of the m-HfO_2 film as a tracer source. Time-of-Flight Secondary Ion Mass Spectrometry (ToF-SIMS) was employed for the subsequent analysis and the obtained isotope profiles were analysed using Finite-Element-Method (FEM) simulations. Cation diffusion in HfO_2 was also studied computationally, where the migration barriers of hafnium ion jumps in m-HfO_2 are determined from DFT calculations with the Climbing-Image Nudged Elastic Band (CI-NEB) method. Molecular Dynamics (MD) simulations with an applied electrical field (for enhancing the mobility of cations) are utilised to obtain migration enthalpies for hafnium diffusion in cubic HfO_2 (c-HfO_2). Defect enthalpies are calculated with static lattice simulations in c-HfO_2 to further the understanding of the results.

5.2 Methodology

5.2.1 Experimental

As noted above, previous studies have shown that silica impurities severely impact the oxygen diffusion behaviour in m-HfO$_2$.[81,95,102,119] For this reason, the industrial standard, silicon wafers, are omitted in this work. While ALD makes use of silicon-free precursors and thus yields silicon-free samples, the film substrates remain as a possible source of silicon, particularly due to silica residues remaining on the surface after polishing. For this reason, the amount of silicon impurities was examined on different substrates [DyScO$_3$, (LaAlO$_3$)$_{0.3}$(Sr$_2$TaAlO$_6$)$_{0.7}$, NdGaO$_3$, SrTiO$_3$, Al$_2$O$_3$ and YSZ] with SIMS. DyScO$_3$ delivered the best results, with the least amount of silicon at the interface of film and substrate. This behaviour is attributed to the self-cleaning effect of perovskite type systems.[120–122] Reactive Ion Beam Etching on the substrate surface followed by HfO$_2$ deposition showed no discernible influence on the amount of silicon. Thus, polished DyScO$_3$ single crystals were used as substrates for the HfO$_2$ thin films.

For the plasma assisted ALD process, tetrakis(ethylmethylamino)hafnium (TEMAH, $\geq$ 99.99 % trace metal basis excluding *ca.* 2000 ppm Zr, SAFC Hitech®) was used as a precursor to deposit HfO$_2$ at $T = 573$ K as a thin, amorphous layer with embedded m-HfO$_2$ nanocrystals of about 3 nm in diameter.[118] An oxygen plasma was used as co-reactant. The ALD process was performed in a FlexAl™ ALD Tool (Oxford Instruments Plasma Technologies, Bristol, United Kingdom). Then, a thin layer of ZrO$_2$ was employed as a diffusion source, deposited by RF-magnetron sputtering (output power of 60 W) from a Zr-target at room temperature under 0.01 mbar pressure in a mixture of 38 sccm Ar and 2 sccm O$_2$. These conditions resulted in a deposition rate of 0.82 nm/min. The aim was to deposit 40 nm of ZrO$_2$ on 100 nm of HfO$_2$.

Diffusion anneals were carried out in a MoSi$_2$ furnace in air inside a closed Al$_2$O$_3$ vessel. The heating rate was 500 K/h for temperatures in the range of $1173 \leq T / K \leq 1323$. Samples were quenched after the anneal by removing them from the MoSi$_2$ furnace and transferring them to another furnace at 873 K. They were then cooled to room temperature with a cooling rate of 500 K/h. This procedure was employed to avoid cracking of the samples through too rapid cooling.

5.2 Methodology

Figure 5.1: SEM image of a ZrO_2/HfO_2 thin film prior to diffusion annealing. Strong charging effects are taking place, indicated by the white spots and the overall blurriness of the image. White markers are exemplary grain size measurements.

Grazing incidence X-ray diffractograms obtained with a X'Pert MRD system (PANalytical, Almelo, Netherlands) indicated the existence of a single monoclinic phase after annealing the sample in air at 1273 K for 20 h. After 20 h at 1373 K, both monoclinic and cubic phase were found. After 10 h at 1423 K, only the cubic phase was found. It is assumed that the phase transition occurs as a result of Dy^{3+} and Sc^{3+} diffusing in from the substrate (detected by ToF-SIMS analysis, see below), similar to the stabilisation of the cubic phase of ZrO_2 upon Y^{3+} doping.[123] To ensure that cation diffusion data refers to the monoclinic phase of HfO_2, only samples annealed at temperatures below 1373 K were subjected to analysis.

Scanning electron microscopy (SEM) images of the thin-film samples prior to annealing were taken with an JEOL JSM-7800F system (JEOL GmbH, Freising, Germany). Results are shown in Fig. 5.1; note the strong charging effects with this system despite a thin layer of platinum on top of the film for better charge equilibration. SEM images taken with a different system after annealing did not show these charging effects. Images after the analysis were taken with an SU8000 system (Hitachi, Tokyo, Japan). From these images it was determined that the grain size varied between 20 nm and 90 nm, depending on the annealing temperature.

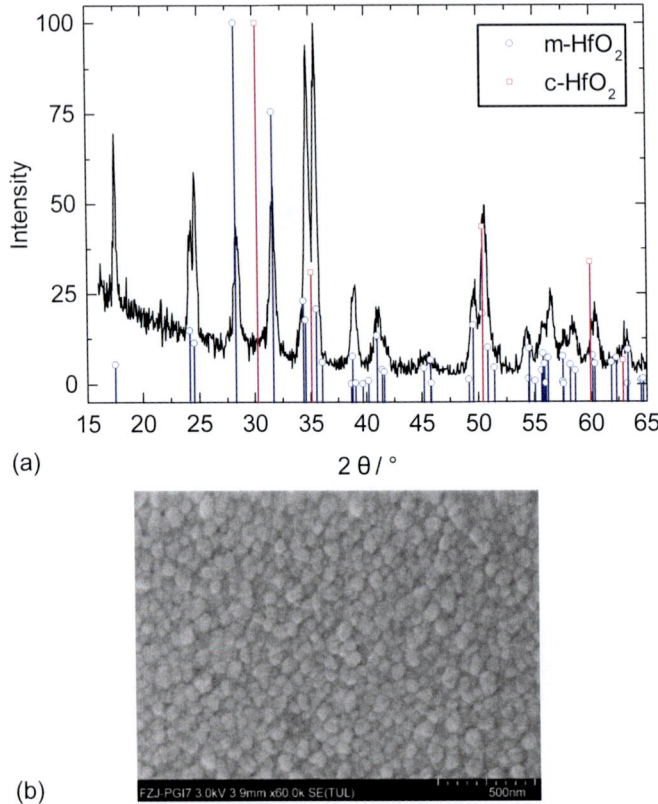

Figure 5.2: (a) Diffractogram of film annealed at 1273 K for 20 h. Magenta lines denote the position of the reflections caused by c-HfO$_2$ and blue lines denote the position of the reflections caused by m-HfO$_2$.[84,124] The black line represents the measured data. No reflections of the cubic phase are evident; (b) SEM image of film annealed at 1273 K.

SEM images and X-ray diffractograms can be found in Fig. 5.2 and 5.3. The different intensities of the experimental diffractograms and the literature values stems from a certain orientation of the thin films during measurement and is inconsequential. Interference microscopy (NT1100, Veeco Instruments Inc., New York, USA) indicated a roughness of $R_q = \pm 2$ nm for the areas relevant to ToF-SIMS analysis (100 µm × 100 µm).

5.2 Methodology

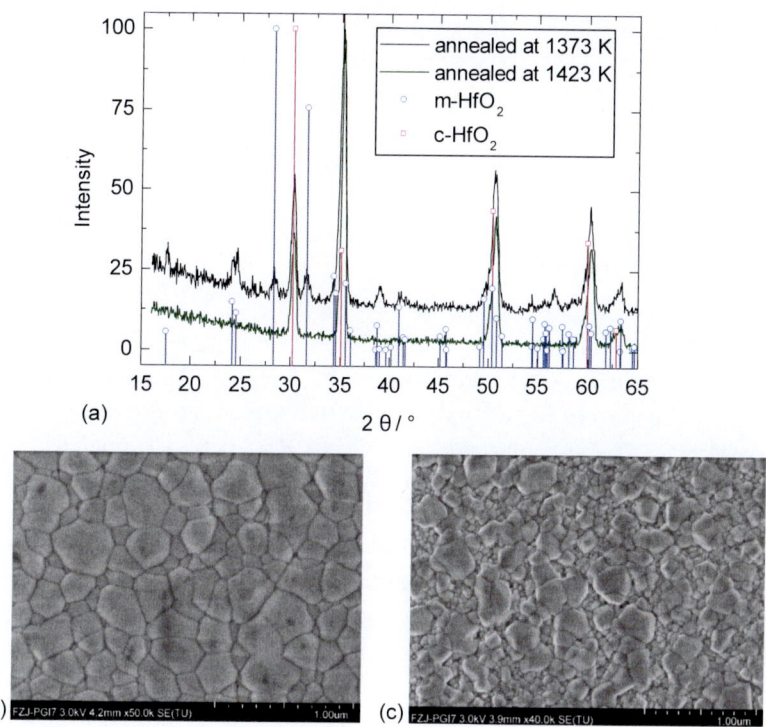

Figure 5.3: (a) Diffractograms of films annealed at 1373 K for 20 h and 1323 K for 10 h, respectively. Magenta lines denote the position of the reflections caused by c-HfO$_2$ and blue lines denote the position of the reflections caused by m-HfO$_2$.[84,124] Results for 1373 K are shifted upwards for better visualisation. Film remains monoclinic at 1373 K, but shows reflections of cubic phase. At 1423 K only cubic phase remains; (b) SEM image of film annealed at 1373 K; (c) SEM scan of film annealed at 1423 K.

ToF-SIMS depth profiles were obtained[52,90] on a TOF.SIMS IV machine equipped with a TOF.SIMS V analyser (ION-TOF GmbH, Münster, Germany). 25 keV Ga$^+$ ions, rastered over an area of 100 µm × 100 µm, were used to generate secondary ions. Preliminary studies indicated that the secondary-ion intensities of AO^- species (A = Hf/Zr) under Cs$^+$ bombardment were higher than those of A^+ species under O_2^+ bombardment. Therefore, 2 keV Cs$^+$ ions were used for sputter etching the sample surface, typically over an area of 400 µm × 400 µm. Measurements were performed with a ToF cycle time of 60 μs in bunched mode,[52] and negative secondary ions were detected. Charge compensation was accomplished with a beam of low-energy (<20 eV) electrons. Secondary-ion intensities of AO^- species were normalised to the intensity of $^{18}O^-$ and to the maximum intensity of the profiles. Crater depths were determined post-analysis by interference microscopy.

5.2.2 Computational

The DFT calculations were performed within the Generalised Gradient Approximation after Perdew, Burke and Ernzerhof.[64] The potentials generated by the Projector Augmented Wave[67] method were used with an energy cutoff of 500 eV. Calculations were performed for a periodic $2 \times 2 \times 2$ supercell of m-HfO$_2$ containing 31 Hf atoms and 64 O atoms (since a Hf^{4+} species was removed, the behaviour of a charged vacancy is considered). k-points were generated by a $2 \times 2 \times 2$ Monkhorst–Pack mesh.[66] For convergence criteria, electronic convergence was set to 1×10^{-4} and ionic convergence to 1×10^{-3}. The CI-NEB[68–70] method was used to determine the energy of the transition state over 3 images along the cation jump. All calculations were implemented in the Vienna Ab initio Simulation Package (VASP).[125,126]

Aside from the DFT calculations, dynamic MD simulations were performed using an NpT ensemble in the temperature range $2500 \leq T \:/\: K \leq 3000$ on a $10 \times 10 \times 10$ supercell of c-HfO$_2$ (12000 ions). A timestep size of 0.001 ps was chosen for all MD simulations. Buckingham-type EPP derived by Lewis and Catlow[76] were employed to describe the short-term interactions up to a range of 6 Å, with the addition of a Coulomb term to describe the long-range interactions up to a range of 20 Å. Table 5.1 shows the EPP parameters used. The Coulomb interactions were computed using a standard Ewald summation with a relative error in forces of 10^{-8}. The same simulations were also performed in the presence of an electrical field in the range of $0.3 \leq E\:/\:\mathrm{MVcm}^{-1} \leq 50$. The simulator achieves this by applying a force to each atom according to $F = qE$.

The ionic charge caused by the removal of 40 Hf^{4+} ions was compensated by decreasing the ionic charge of all oxygen ions slightly (−1.98 instead of −2). The Large-scale Atomic/Molecular Massively Parallel Simulator (LAMMPS)[127] was employed for the dynamic simulations.

5.2 Methodology

Table 5.1: Empirical pair potential parameters derived from ref. 76.

	A / eV	ρ / Å	C / eV·Å^6
Hf^{4+}–O^{2-}	1454.6000	0.3500	0.0000
O^{2-}–O^{2-}	22764.0000	0.1490	20.3700

The General Utility Lattice Program (GULP)[128] was used for static Mott–Littleton type defect simulations in c-HfO$_2$.[129] Again, Buckingham-type EPP derived by Lewis and Catlow[76] (see table 5.1) were employed to describe the short-term interactions up to a range of 6 Å. Standard Coulomb terms described the long-range interactions. A defect centre in the middle of the simple unit cell and region size radii of 14 Å and 22 Å for the Mott–Littleton type defect simulations were chosen. In the first spherical region, all particle interactions are calculated explicitly while in the second spherical region, interaction energies are approximated. Two hafnium vacancies and one fixed hafnium interstitial were inserted to probe for the ideal migration path in c-HfO$_2$. Additional oxygen vacancies were introduced to investigate their effect on the migration path.

5.3 Results

5.3.1 Experimental

The sample geometry and the investigated length-scale of the diffusion samples are indicated in Fig. 5.4(a). Fig. 5.4(b) shows the intensity profiles for ZrO⁻ and HfO⁻ secondary ions across the structure obtained prior to the diffusion anneal by means of SIMS depth profiling. The two different layers are well-defined and homogeneous. Examining the interface region, it becomes apparent that the range of depths over which the ZrO⁻ signal decreases is approximately equal to the range of depths over which the HfO⁻ signal increases. This suggests that the broadening of the signals comes from the samples, *i.e.* it is due to the roughness of the ZrO$_2$ surface and/or the ZrO$_2$|HfO$_2$ interface. If direct recoil and ion beam mixing (SIMS effects) were responsible, the two ranges of depths would not be similar. At a depth of *ca.* 140 nm the hafnia layer ends and the substrate layer begins, which is the expected result according to the process specifications.

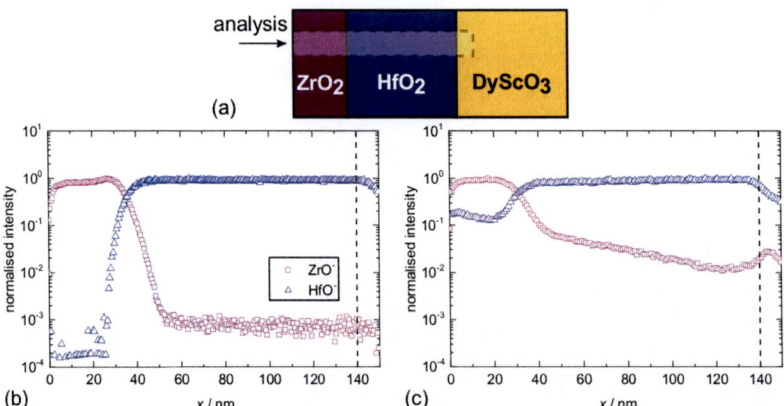

Figure 5.4: (a) Sketch of sample geometry and the analysed region; (b) Normalised intensities of ZrO⁻ and HfO⁻ secondary ions obtained by ToF-SIMS depth profiling for a sample prior to the diffusion anneal; Vertical dashed line indicates the interface with the DyScO$_3$ substrate; (c) Normalised intensities of ZrO⁻ and HfO⁻ secondary ions obtained by ToF-SIMS depth profiling for a sample annealed at $T = 1273\,\text{K}$ for $t = 20\,\text{h}$.

5.3 Results

After diffusion annealing, the intensity profile of ZrO⁻, as shown in Fig. 5.4(c), still displays a constant plateau in the ZrO_2 layer; at the $ZrO_2|HfO_2$ interface, the broadening is less steep than in the sample prior to the diffusion anneal, suggesting cation diffusion; and a long profile, extending from ≈ 45 nm up to the substrate at ≈ 135 nm, has developed. The position of the $HfO_2|DyScO_3$ interface has apparently shifted slightly, but careful examination revealed that it is actually a shift of the $ZrO_2|HfO_2$ interface. The thickness of the HfO_2 layer has not changed. This shift is ascribed to the densification of the ZrO_2 layer upon thermal treatment (since it was sputtered at room temperature) and a consequent reduction in film thickness. Near the substrate an upturn in the ZrO⁻ intensity profile is observed. This upturn is attributed to a SIMS artefact that arises as the sputter front passes through the interface from the HfO_2 film to the $DyScO_3$ substrate. Simulations of diffusion (see below) did not reproduce this upturn. The obvious interpretation of the profile is that the broadening is due to diffusion in bulk m-HfO_2 and the long profile is due to a combination of fast grain-boundary diffusion and slow diffusion out of the boundary. Given that the area of the ToF-SIMS analysis was $100\,\mu m^2$ and the area of a single grain is of the order of $10^3\,nm^2$, the profile refers to the average over $\sim 10^5$ grains and their grain boundaries. The diffusion profile obtained for HfO⁻ shows similar, though not identical, behaviour. Going from the substrate towards the surface, a constant plateau of HfO⁻ in the HfO_2 layer is seen; then a decrease corresponding to diffusion and the broadening arising from interfacial roughness; but then an increase in HfO⁻ intensity towards the surface, suggesting surface diffusion of HfO_2 from the film to the external ZrO_2 surface, and diffusion inwards from this surface. In the following the HfO⁻ profiles are not considered further, as the focus of this work lies on the ZrO⁻ profiles in HfO_2.

Given the complexity of the diffusion problem — diffusion along two different paths (bulk and grain boundary) in a medium of finite extent from a non-trivial initial distribution — the method of choice is to solve the diffusion equation numerically in two dimensions in order to obtain the relevant diffusion coefficients. To this end, Finite-Element-Method (FEM) simulations were implemented in COMSOL Multiphysics® (COMSOL AB, Stockholm, Sweden) for a simulation box of length l_{film} and of width $d_{gr} + d_{gb}$ (of which d_{gb} is the region within which fast grain-boundary diffusion takes place). Bulk and grain-boundary diffusion coefficients were assumed to be isotropic and the same for both ZrO_2 and HfO_2 layers. The assumption of isotropic diffusion coefficients for non-isotropic m-HfO_2 is reasonable, since each diffusion profile is an average over $\sim 10^5$ grains

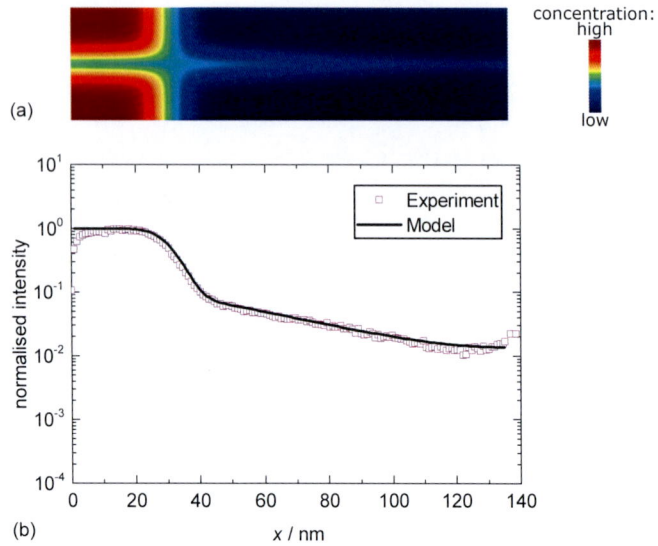

Figure 5.5: Continuum simulations of zirconium diffusion into a polycrystalline thin-film at $T = 1273\,\text{K}$ for $t = 20\,\text{h}$. (a) 2D concentration heat map of the simulation cell, with the grain boundary running through the centre of the cell; (b) Comparison of simulated and experimental diffusion profiles.

(and the grains show no preferred orientation according to XRD scans). In the simulations l_{film} was set to the film's thickness, d_{gb} was chosen to be $1\,\text{nm}$, and d_{gr} was assumed to be constant at each temperature (*i.e.* grain growth was assumed to take place faster than diffusion into the grains), with values taken from the SEM images. The intensity profile prior to the diffusion anneal [*e.g.* Fig. 5.4(b)] was used as the initial condition.

Fig. 5.5(a) shows a 2D concentration heat map obtained from an FEM simulation of zirconium diffusion in the layer structure. The different regions and grain boundaries of the model are clearly visible in the heat map. The concentration heat map was averaged parallel to the surface to obtain a simulated diffusion profile that was compared visually with the experimental data. D_{b}, D_{gb} and d_{gr} were varied until good agreement was found, as shown in Fig. 5.5(b). (*N.b.*: only small adjustments to d_{gr} from the SEM values were necessary.) Unlike the standard analysis of grain-boundary diffusion,[36,37] which yields the grain-boundary

5.3 Results

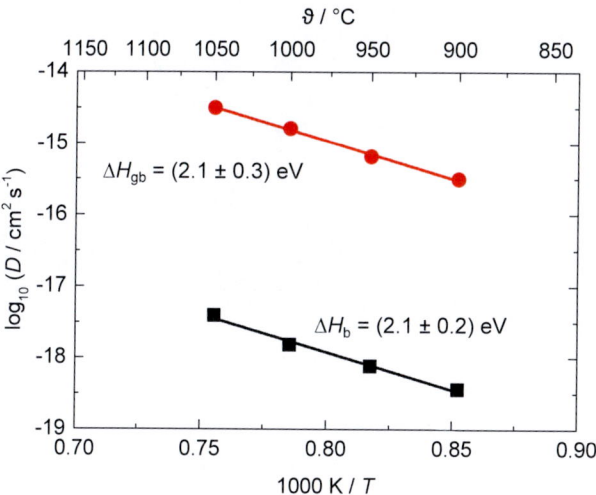

Figure 5.6: Diffusion coefficients of zirconium in bulk m-HfO₂ and along its grain boundaries as a function of inverse temperature.

diffusion product $d_{gb}D_{gb}$, these simulations give the grain-boundary diffusion coefficient D_{gb}, since d_{gb} is explicitly specified. This means that the values of D_{gb} are specific to the value of d_{gb}; if accelerated diffusion takes place within a region wider or narrower than $d_{gb} = 1$ nm, then the values obtained for D_{gb} will change (see later). Diffusion coefficients of zirconium obtained for the m-HfO₂ films as a function of annealing temperature are plotted in Fig. 5.6. D_{gb} is seen to be approximately 3 orders of magnitude higher than D_b at all temperatures. For bulk diffusion of zirconium in m-HfO₂, an activation enthalpy of $\Delta H_b = (2.1 \pm 0.2)$ eV is obtained. The activation enthalpy of zirconium diffusion in the grain boundary is the same at $\Delta H_{gb} = (2.1 \pm 0.3)$ eV.

5.3.2 Computational

The assumed diffusion mechanisms of oxygen in c-HfO₂ and m-HfO₂ are described in detail in chapter 2.3.1. In m-HfO₂, due to the monoclinic distortion, hafnium is sevenfold coordinated by oxygen and two different oxygen sites exist, half of which are threefold and half fourfold coordinated. Although the num-

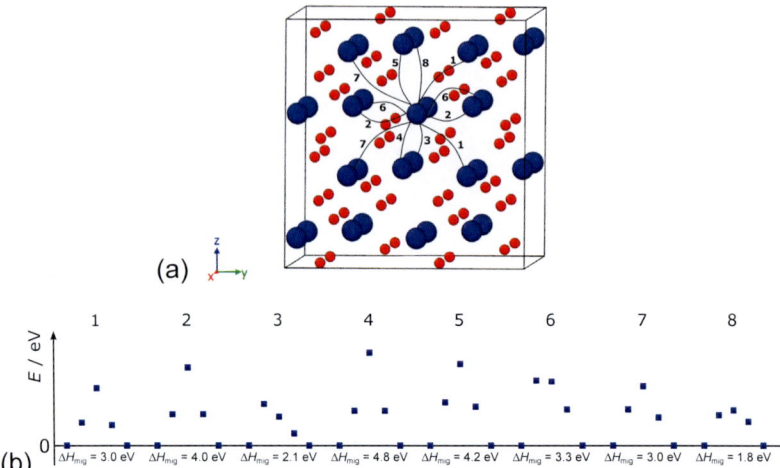

Figure 5.7: (a) Crystal structure of m-HfO$_2$, indicating the eight symmetry-inequivalent jumps of a hafnium cation by a vacancy mechanism. Hafnium ions are shown in blue, oxide ions in red. Lines do not indicate exact jump paths. Figure created with OVITO.[130] (b) Schematic energy landscapes of the different jumps in m-HfO$_2$ showing the migration barriers. Each jump consists of 3 images and start/end-positions. The images are not equidistant. The y-axis does not display absolute energy values, as only the relative values are important.

ber of next-nearest neighbouring hafnium ions and thus the number of possible jumps remains the same as in c-AO$_2$, many of these jumps are no longer symmetry equivalent. In fact, only four of these jumps are symmetry equivalent, leaving eight distinct possible jumps (see Fig. 5.7(a)), in contrast to the single, unique jump of the cubic structure. CI-NEB calculations were used to calculate the migration barriers by a vacancy mechanism for these hafnium ion jumps in the monoclinic structure. The results are plotted in Fig. 5.7(b).

The activation barriers are observed to vary from 1.8 eV to 4.8 eV. This range of values does not seem to be consistent with experiments. One critical point, however, is the combinations of jumps that are required so that hafnium can migrate through the cell from one side to the other. In this regard, jumps 3 and 8, with 2.1 eV and 1.8 eV, provide together a path through the cell in the ⟨001⟩ direction. The second critical point concerns the jump rates at the temperatures of the exper-

5.3 Results

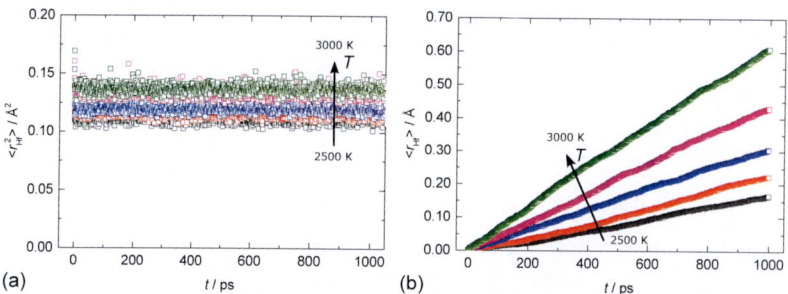

Figure 5.8: Mean displacements of Hf^{4+} ions during MD simulations. Temperatures are 2500 K, 2625 K, 2750 K, 2875 K and 3000 K. (a) $\langle r_{\text{Hf}}^2 \rangle$ of Hf^{4+} ions over time; (b) $\langle r_{\text{Hf}} \rangle$ over time with an applied electrical field of 7 MV/cm in $\langle 011 \rangle$ direction.

iments, $\Gamma_v = \nu_0 \exp(\Delta S_{\text{mig,v}}/k_B)\exp(-\Delta H_{\text{mig,v}}/k_B T)$. Assuming that the attempt frequencies (ν_0) and the activation entropies of migration ($\Delta S_{\text{mig,v}}$) do not vary much for the various jumps, one finds that, at $T = 1100$ K, jumps over the other barriers occur far less frequently in comparison (for 3 eV, a factor of 10^4 less; for 4.8 eV, a factor of 10^{12} less). Hence, hafnium migration in other directions will not be observed at the temperatures of the experiments because those paths have substantially higher barriers. It is thus concluded that the activation energy of hafnium-vacancy migration at the temperatures of interest in a polycrystal with randomly oriented grains will be given by the (largest) migration barrier in the $\langle 001 \rangle$ direction, *ca.* 2 eV.

To gain additional information on the cation migration mechanisms, MD simulations were performed on hafnium diffusion in c-HfO$_2$. Simulations were not performed in m-HfO$_2$ because the EPP, while suitable for oxygen diffusion in m-HfO$_2$,[29] did not reproduce the monoclinic cell reliably in the presence of cation vacancies at the temperatures of interest. Since comparable AO_2 oxides crystallise predominantly in the cubic structure and the cubic phase is the thermodynamically favoured phase at the temperatures investigated, it is not unreasonable to look at c-HfO$_2$.

Mean square displacements $\langle r_{\text{Hf}}^2 \rangle$ obtained from molecular dynamic simulations as a function of time t are shown in Fig. 5.8(a) for a c-HfO$_2$ system containing 1 % hafnium vacancies. No increase in $\langle r_{\text{Hf}}^2 \rangle$ can be observed at any of the investigated

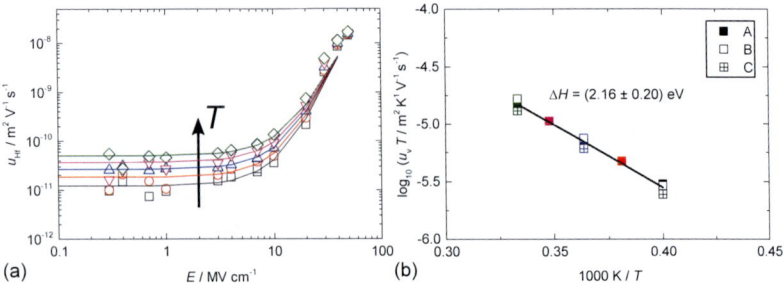

Figure 5.9: (a) Mobilities of Hf^{4+} as a function of the external electrical field in $\langle 011 \rangle$ direction. (b) Arrhenius plot for hafnium vacancy mobilities obtained from extrapolation of mobilities in (a). (A) field in $\langle 011 \rangle$ direction; (B) field in $\langle 100 \rangle$ direction; (C) 0.5 % vacancies instead of 1 % vacancies, field in $\langle 011 \rangle$ direction.

temperatures; ergo, no hafnium diffusion has taken place in the investigated time span.

In order to measure ion migration for hafnium without increasing the temperature to experimentally unreasonable values, different electrical fields were applied to increase the mobility of ions in the given temperature range and extract the mean displacements ($\langle r_{Hf} \rangle$) of the cations (see Fig. 5.8(b)). The drift velocity v_d of the Hf^{4+} ions is then equal to the gradient of $\langle r_{Hf} \rangle$ versus t. From that, the ionic mobility u_i can be obtained. Fig. 5.9(a) shows the mobilities in dependence of the applied electrical field.

By extrapolating the mobilities in the linear (small-field) region of Fig. 5.9(a) to an electrical field of $0\,\mathrm{MV/cm}$, the field-independent mobilities are obtained. This extrapolation is done by fitting the mobility data with Eq. 2.34:

$$u_i E^{FS} = v_d = B \cdot \left[\exp\left(-\frac{\Delta H^f_{mig}}{k_B T}\right) - \exp\left(-\frac{\Delta H^r_{mig}}{k_B T}\right) \right], \quad \text{(2.34 revisited)}$$

with $B = 2\left(1 - n_i\right) a_i \nu_0 \exp\left(\frac{\Delta S_{mig}}{k_B}\right)$.

5.3 Results

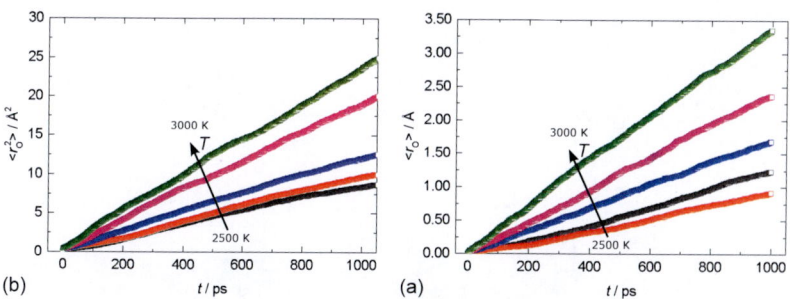

Figure 5.10: (a) $\langle r_O^2 \rangle$ of O^{2-} ions over time. (b) $\langle r_O \rangle$ over time with an applied electrical field of $7\,\mathrm{MV/cm}$ in $\langle 011 \rangle$ direction. Temperatures are $2500\,\mathrm{K}$, $2625\,\mathrm{K}$, $2750\,\mathrm{K}$, $2875\,\mathrm{K}$ and $3000\,\mathrm{K}$.

The required migration barrier in direction of the field is calculated according to Eq. 2.37:

$$\Delta H_{\mathrm{mig}}^{\mathrm{f/r}} = H\left(x_{\mathrm{max}}^{\mathrm{f/r}}\right) - H\left(x_{\mathrm{min}}\right)$$
$$= \Delta H_{\mathrm{mig}}\left[\sqrt{1-\gamma^2} \mp \gamma\left(\frac{\pi}{2}\right) + \gamma \arcsin\gamma\right], \quad (2.37 \text{ revisited})$$

with $\gamma = \frac{(2|z_i|eE^{\mathrm{FS}}a_i)}{(\pi \Delta H_{\mathrm{mig}})}$. See chapter 2.4.1 for more in-depth information.

The extrapolated values for an electrical field of zero are shown in an Arrhenius plot in Fig. 5.9(b) for the hafnium vacancy mobilities. They have been obtained for different field directions and differing numbers of cation vacancies to increase statistical accuracy. From the slope a migration enthalpy of $\Delta H_{\mathrm{mig}} = (2.2 \pm 0.2)\,\mathrm{eV}$ (where the error corresponds to twice the standard deviation) can be extracted.

Fig. 5.10(a) shows the mean square displacement of oxygen ions during the MD runs. Contrary to the hafnium ions, oxygen displays significant movement, even looking at the non-field-enhanced simulations. The investigated system is created without any oxygen vacancies. This phenomenon will be discussed in relation to the other simulations in chapter 5.4.1.

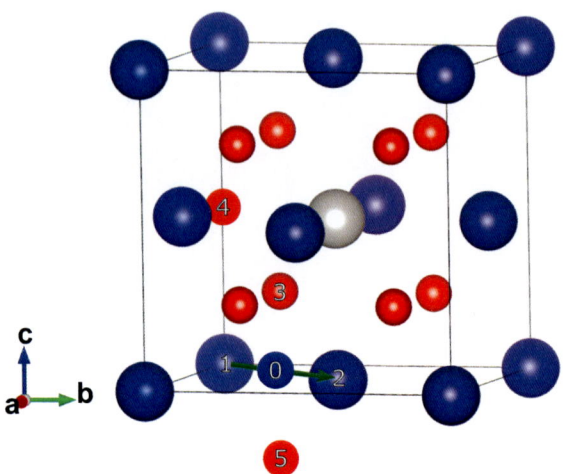

Figure 5.11: Unit cell and relevant ions for migration path simulations with GULP. Hafnium ions are shown in blue, oxide ions in red. The defect centre is shown in grey. The numbers index the ion positions.

To further investigate the diffusion path of Hf^{4+} in c-HfO_2, static simulations were performed. NEB simulations done in LAMMPS showed a migration barrier of $\Delta H_{mig} = 6.63\,eV$. The chosen path is slightly curved in the $\langle 110 \rangle$ plane. Studies on similar AO_2 systems have observed similar migration paths.[25,26] In order to complement the NEB simulations, static lattice simulations were performed in the GULP package (also see section 5.2.1). These simulations were set up by defining a defect centre as part of a Mott–Littleton type defect calculation and removing certain ions, determining a defect energy in the process. Fig. 5.11 shows the simulation cell input and illustrates the ions removed in the process of simulating migration paths. The grey defect centre is not an ion and serves as a point of reference that stays constant throughout all static lattice simulations.

First, the hafnium ions marked as 1 and 2 in Fig. 5.11 are removed. Then, an interstitial hafnium ion is inserted between positions 1 and 2 along the green path. The structure is then force-optimised and a defect energy of $\Delta H_{mig,static} = 6.69\,eV$ is obtained for the interstitial hafnium ion. The path is slightly curved in the $\langle 110 \rangle$ plane. This is a different approach compared to the aforementioned NEB

5.3 Results

simulations in LAMMPS, but it produces a similar result. Introducing an oxygen vacancy next to this path, at position 3, leads to a slight increase of the migration barrier, up to 6.95 eV, and a straight jump path that is slightly elevated in $\langle 001 \rangle$ direction. Introducing an oxygen interstitial at position 4 lowers the migration barrier to 5.93 eV and shifts the migration path in the $\langle 110 \rangle$ plane, but to a lesser degree than the path in the cell with the intact oxygen sublattice. A hafnium jump in the vicinity of two oxygen vacancies at positions 3 and 5 displays a higher migration barrier of 8.75 eV and follows a straight line from position 1 to 2. Additional defect energies for an oxygen interstitial (position 4), an oxygen vacancy (position 3), a hafnium vacancy (position 1), an anti-Frenkel defect, and combinations of them are shown in table 5.2.

Table 5.2: Defect energies and distances for different defects and defect pairs in c-HfO$_2$. Number in brackets is corrected by the defect energy of a single hafnium vacancy.

Defect type	0 V_{Hf}/eV	Distance/Å
O_i''	-8.76	//
$V_O^{\bullet\bullet}$	16.66	//
$V_O^{\bullet\bullet} - O_i''$	6.19	2.21
V_{Hf}''''	90.26	//
$O_i'' - V_{Hf}''''$	84.03 (-6.23)	2.55
$V_O^{\bullet\bullet} - V_{Hf}''''$	103.14 (12.88)	2.21
$(V_O^{\bullet\bullet} - O_i'') - V_{Hf}''''$	100.91 (10.64)	2.12

The sum of the oxygen interstitial defect energy at infinite dilution and oxygen vacancy defect energy at infinite dilution is 7.91 eV (with a defect separation of 2.21 Å). Comparatively, the formation of an anti-Frenkel defect is energetically favoured by 1.72 eV with a defect energy of 6.19 eV, at the same defect separation. In the vicinity of a hafnium vacancy, the defect energies change. The energy of the hafnium vacancy is subtracted, so that the numbers remain comparable. The sum of the two separate and infinitely diluted defects is reduced to 6.64 eV, while the anti-Frenkel energy is increased to 10.64 eV. The association energy E_a of an oxygen vacancy and a hafnium vacancy can then be calculated according to

$$E_a = E_{tot}(V_O^{\bullet\bullet}) + E_{tot}(V_{Hf}'''') - E_{tot}(V_O^{\bullet\bullet} - V_{Hf}'''') = -3.79 \, \text{eV}. \tag{5.3}$$

5.4 Discussion

5.4.1 Bulk diffusion

The good agreement between ΔH_D obtained experimentally and $\Delta H_{\text{mig,v}}$ obtained computationally for cation transport in m-HfO$_2$ strongly suggests ([see Eq. 5.1) that $\Delta H_{\text{gen,v}}$ for the investigated non-equilibrium samples was zero and hence that the concentration of hafnium vacancies does not vary with temperature, in accord with the initial premise (see section 5). The investigation of non-equilibrium samples thus opens a new avenue to studying the migration of slow-moving minority defects.

Compared with values obtained computationally for cation-vacancy migration in other AO_2 systems — $\Delta H_{\text{mig,v}}$ is over $4\,\text{eV}$ for CeO$_2$[26] and over $5\,\text{eV}$ for UO$_2$[27,28,131] and ZrO$_2$[132] —, $\Delta H_{\text{mig,v}}$ for cations in m-HfO$_2$ has similar values ($3\,\text{eV}$ to $5\,\text{eV}$) but also two substantially lower ($\approx 2\,\text{eV}$) ones according to the DFT calculations. The difference between these three oxides and HfO$_2$ is the crystal symmetry: CeO$_2$ and UO$_2$ adopt the high-symmetry cubic form (and ZrO$_2$ was considered in the computer simulations[132] as a hypothetical cubic structure), whereas HfO$_2$ has a strongly distorted monoclinic structure.

Metlenko et al.[133] proposed a general rule for the effect of structural perturbations on ion migration. They hypothesised that, for ions that are highly mobile in a given structure, structural perturbations lead to a decrease in ion mobility, whereas for ions that are immobile, structural perturbations lead to an increase. Indeed, isothermal rates of oxide-ion transport in monoclinic and tetragonal ZrO$_2$ and HfO$_2$ are found to be much lower than in the cubic form.[29,134–136] One would expect, therefore, the strong structural perturbation of m-HfO$_2$ to lead to higher rates of cation diffusion than in the cubic counterparts. In the present case, one could argue more specifically in terms of the oxygen coordination of the cations. A sevenfold coordination of hafnium by oxygen increases their mobility compared to that in an eightfold coordination; for CeO$_2$ and UO$_2$ such an enhancement of cation mobility is found in the presence of oxygen vacancies[137,138] (specifically, an increase in grain-boundary mobility was observed in an oxygen-poor atmosphere). If oxygen vacancies in a cubic fluorite lattice are understood as a perturbation of the eightfold coordination of the A cation towards a sevenfold

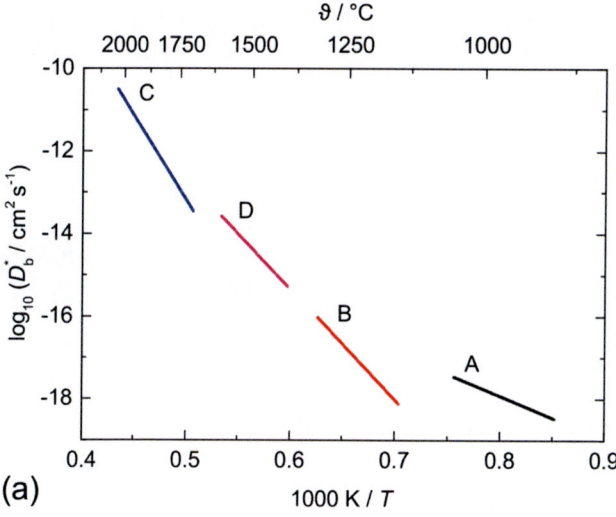

Figure 5.12: Comparison of diffusion coefficients for cation bulk diffusion in selected fluorite-based oxides: (A) Zr in m-HfO$_2$, this study; (B) Zr in Gd-doped (0.5 %) CeO$_2$;[38] (C) Hf in Er$_2$O$_3$-stabilized (10 %) HfO$_2$;[139] (D) Hf in 3 mol% Y$_2$O$_3$-stabilized ZrO$_2$.[140]

coordination, then the sevenfold coordination of hafnium in the monoclinic crystal structure might have a similar increasing effect on cation mobility and explain the low migration enthalpies.

Another benefit of this approach is that it allows the estimation of n_v, the site fraction of cation vacancies in the investigated samples. Since the bulk diffusion coefficients can be expressed in terms of n_v, a jump distance a_{jump} and the jump rate Γ_v,

$$D_b = n_v a_{jump}^2 \Gamma_v, \qquad (5.4)$$

one finds, with $a_{jump} = 0.36\,\text{nm}$, $\nu_0 = 10^{13}\,\text{s}^{-1}$, and $\Delta S_{mig,v}/k_B = 0$, $n_v \sim 10^{-8}$ (see Eq. 2.22 and 2.23). These values are consistent with cation vacancies being minority defects in acceptor-doped AO_2 systems.

In Fig. 5.12 the bulk cation diffusion coefficients determined in this work are compared with selected experimental data sets of similar fluorite-structured systems reported in the literature. If the datasets B, C and D are considered together and the data is extrapolated to the temperatures of the experiments done here, one finds that the cation bulk diffusivities determined here are significantly higher than the ones found in those systems and the activation enthalpy is significantly lower at $\Delta H_b = (2.1 \pm 0.2)\,\text{eV}$. Beschnitt and De Souza[38] reported $\Delta H_b = (5.5 \pm 0.4)\,\text{eV}$ for Zr diffusion in Gd-doped (0.5 %) CeO_2 (cubic); Tesch et al.[139], $\Delta H_b = (8.0 \pm 0.3)\,\text{eV}$ for Hf diffusion in Er_2O_3-stabilized (10 %) HfO_2 (cubic); and Swaroop et al.[140], $\Delta H_b = (5.3 \pm 0.9)\,\text{eV}$ for hafnium diffusion in $3\,\text{mol}\%$ Y_2O_3-stabilized ZrO_2 (tetragonal). The difference is attributed to the samples in this work having a higher site fraction of cation vacancies and a lower activation barrier of migration (being monoclinic rather than cubic or tetragonal).

5.4.2 Molecular Dynamics Simulations

The field-enhanced MD simulations agree well with both the experimental results and the DFT results. Almost the same migration enthalpy of $\Delta H_{mig} \approx 2.2\,\text{eV}$ is obtained, even though the dynamic simulations investigated the cubic structure. Curiously, molecular static simulations with NEB and lattice simulations obtained a significantly different migration barrier of $\Delta H_{mig} \approx 6.6\,\text{eV}$, which is more in line with other cubic AO_2 systems than the monoclinic HfO_2 phase.[25–28] The migration path found for those systems also strongly resembles the curved path found for c-HfO_2 here.

Subsequently, the question arises whether the cation migration in c-HfO_2 simply follows the same behaviour as other cubic AO_2 systems, making the results of the field-enhanced MD simulations the outlier; or if there is some detail which is not taken into consideration by the molecular static simulations, but is included in field-enhanced MD simulations, DFT calculations and experiment.

Significant oxygen migration has been reported for the simulations without and with field. This is somewhat surprising, since no oxygen vacancies were introduced to the simulation cells. Trajectories of O^{2-} from the MD run indicate the creation of short-lived oxygen vacancies near hafnium vacancies.[132] An oxygen ion moves out of its designated place in the lattice and moves back in after only a few timesteps. Dynamic processes such as a cooperative migration mechanism of hafnium and oxygen vacancies would be not reproducible by molecular static NEB simulations, possibly explaining the differences between experimentally

5.4 Discussion

and dynamically observed activation enthalpy of diffusion and the statically predicted migration barrier. The anti-Frenkel energy is relatively high at 6.19 eV and the static lattice calculations indicate that it increases in the vicinity of hafnium vacancies. This makes the stated hypothesis that short-lived oxygen vacancy and oxygen interstitial pairs are created unlikely. On the other hand, the migration barrier in the vicinity of an oxygen interstitial is slightly reduced, according to the defect simulations, which supports the hypothesis, but no migration barrier as low as $\Delta H_{\text{mig}} \approx 2.2\,\text{eV}$ could be found. Subsequently, the static lattice calculations remain inconclusive. A local perturbation of the cubic structure towards the monoclinic structure is also imaginable as the reason for a lower migration barrier in the dynamic simulations. This would not be reproduced by the static calculations either.

To try and investigate the issue quantitatively, a look at Eq. 5.1 is required: ΔH_D is the result obtained by experiments and generally ΔH_{gen} needs to be taken into account in order to obtain the important ΔH_{mig}. For both experimental and computational case ΔH_{gen} should be zero. For the experimental samples ΔH_{gen} has to be zero due to the non-equilibrium state and for the simulations all defect concentrations are set by using an NpT ensemble, meaning that ΔH_{gen} is guarenteed to be zero. Thus, according to Eq. 5.1, ΔH_{mig} should be equal to ΔH_D and both field-enhanced MD and experiment give similar results. The molecular static simulations give different, much higher results, therefore something is missing in the considerations. A possible explanation would be that the molecular static simulations do in fact give the correct ΔH_{mig}, and that ΔH_{gen} for the system is negative, therefore resulting in a ΔH_D that matches the results of the other methods.

Oxygen ion transport is observed during the dynamic simulations with and without field and therefore it is assumed that an associate of hafnium vacancy (V_{Hf}'''') and oxygen vacancy ($V_O^{\bullet\bullet}$) forms, which is neglected in the molecular static simulations. In that case, ΔH_{gen} needs to contain the formation enthalpy of the associate ΔH_a. The law of mass action for the associate $V_O^{\bullet\bullet} - V_{\text{Hf}}''''$ is defined as:

$$K_a = \frac{[V_O^{\bullet\bullet} - V_{\text{Hf}}'''']}{[V_O^{\bullet\bullet}][V_{\text{Hf}}'''']} = \exp\left(\frac{\Delta S_a}{k_B}\right)\exp\left(\frac{-\Delta H_a}{k_B T}\right) = \exp\left(\frac{-\Delta G_a}{k_B T}\right). \quad (5.5)$$

$[V_{\text{Hf}}'''']$ is constant for both experiment and simulation and can be consolidated into K_a. $[V_O^{\bullet\bullet}]$ is constant in the experiments due to being set by acceptors, but might be created through anti-Frenkel defects in the simulations, which then also have to be taken into consideration. $[V_O^{\bullet\bullet}]$ and $[O_i'']$ can be expressed through their own law of mass action:

$$K_{a-Fr}^{\frac{1}{2}} = [V_O^{\bullet\bullet}] = [O_i''] = \exp\left(\frac{\Delta S_{a-Fr}}{2k_B}\right) \exp\left(\frac{-\Delta H_{a-Fr}}{2k_BT}\right). \tag{5.6}$$

The entropy terms can be neglected, since they are generally assumed to be small, so that $\exp(\Delta S/k_B) \approx 1$. By inserting Eq. 5.6 into Eq. 5.5, an expression describing the site fraction of the associate is formed:

$$K_a K_{a-Fr}^{\frac{1}{2}} = [V_O^{\bullet\bullet} - V_{Hf}''''] = \exp\left(\frac{-\Delta H_a}{k_BT}\right) \exp\left(\frac{-\Delta H_{a-Fr}}{2k_BT}\right). \tag{5.7}$$

As seen in the static lattice calculations, ΔH_{a-Fr} in c-HfO$_2$ is ca. 6 eV, depending on the distance of vacancy and interstitial.[141,142] To describe the enthalpy of defect generation ΔH_{gen}, the concentrations of the defects need to be inserted into Eq. 5.2. Since the enthalpies are assumed to be invariant with temperature (reasonable, since static simulations are investigated) the following holds true for the enthalpy of defect generation:

$$\Delta H_{gen} = \Delta H_a + \frac{1}{2}\Delta H_{a-Fr}. \tag{5.8}$$

With $\frac{1}{2}\Delta H_{a-Fr} \approx 3$ eV and $\Delta H_a \approx -3.8$ eV, ΔH_{gen} is -0.8 eV. This is not sufficient to explain the different results for the molecular static simulations of 6.6 eV and the dynamic simulations of 2.2 eV. Further simulations should be performed to completely understand the difference between the field-enhanced MD simulations and the static simulations.

5.4.3 Grain-Boundary Diffusion

Fast grain-boundary diffusion of cations has also been previously observed in cubic AO_2 type oxides.[38,140] In some cases, *e.g.* ref. 38, the ratio of the activation enthalpies was also found to be $r = \Delta H_{gb}/\Delta H_b \approx 1$, rather than the expected[33–37] $r \approx 0.5$. This would suggest that the traditional picture of fast grain-boundary diffusion taking place along the grain-boundary core does not hold here (as it requires $r \approx 0.5$). The alternative possibility is fast diffusion along space-charge layers. And the presence of such layers at grain boundaries is well established for various acceptor-doped AO_2-type oxides,[40–45] in which positive grain-boundaries are compensated by negative space-charge layers, in which (positive) oxygen vacancies $c_{v,o}$ are depleted and (negative) acceptor dopants c_{acc} are accumulated. Fig. 5.13 shows the respective concentrations in equilibrium. The important point in this case is that the cation vacancies $c_{v,a}$, as negatively charged defects, are also accumulated in the space-charge layers. It is this accumulation that gives rise to the enhanced rates of cation diffusion.[39]

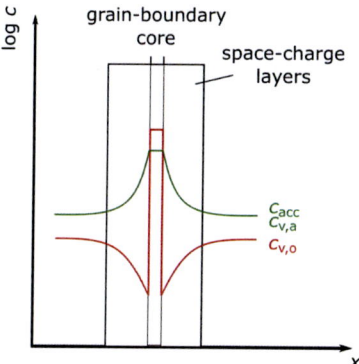

Figure 5.13: Equilibrium defect concentrations in an acceptor-doped AO_2 system. Displayed are positive oxygen vacancies $c_{v,o}$, negative acceptor dopants c_{acc} and negative cation vacancies $c_{v,a}$.

Fig. 5.14 shows that this picture is consistent with the experimental data, and the space-charge potential is extracted. The comparison is made on the basis of the grain-boundary diffusion product, $D_{gb}d_{gb}$, for three reasons. First, it is the quantity that is usually extracted from diffusion experiments in Harrison type B kinetics. Second, it avoids the arbitrary specification of d_{gb} used in the experimental section. Third, it is the quantity that comes out of the space-charge analysis.[39]

Incidentally, one can easily discount the alternative explanation, that the observed behaviour is simply due to more cation vacancies being present in the grain-boundary core. This ignores that space-charge layers are present at grain boundaries in acceptor-doped AO_2 systems. In addition, it requires the migration barrier for cation vacancies at the boundary to be the same as in the bulk: this is extremely unlikely. And it requires the cation vacancies to be neutral (if they were charged the question of space-charge zones raises its head). Measurements of leakage currents and subsequent analysis in line with the discussion of Gritsenko et al.[143] give a range for the concentration of acceptor concentrations c_{acc} in the HfO$_2$ thin films, which amounts to $10^{19}\,\text{cm}^{-3}$ to $10^{20}\,\text{cm}^{-3}$. Using a range of $\epsilon_r = 15 - 25$ for the relative dielectric permittivity of m-HfO$_2$,[4,20,21,144] a found space-charge potential of $\Phi_0 \approx 0.82\,\text{V}$ is consistent with the experimental data (see Fig. 5.14). The impact of the range in acceptor concentration and dielectric permittance is rather small and the space-charge potential is neither unreasonably high nor low.[40,50]

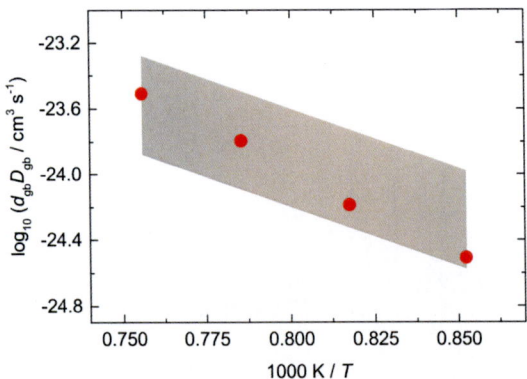

Figure 5.14: Comparison of $D_{\text{gb}}d_{\text{gb}}$ values obtained experimentally (red circles) and predicted from a space-charge potential of $\Phi_0 \approx 0.82\,\text{V}$ (grey area). The predicted range of $D_{\text{gb}}d_{\text{gb}}$ is obtained for $10^{19} \leq c_{\text{acc}}/\text{cm}^{-3} \leq 10^{20}$ and $15 \leq \epsilon_r \leq 25$.

5.5 Summary

Studying experimentally the diffusion of 'immobile' minority species, widely considered as very challenging, is achieved by utilising a low-temperature preparation method, ALD, to prepare non-equilibrium samples in which the concentration of the minority defects is constant. The behaviour of these defects was probed subsequently by performing cation diffusion experiments. The measured diffusion profiles display two features, are analysed by solving the diffusion equation numerically, and yield bulk diffusion coefficients and grain-boundary diffusion coefficients.

The activation enthalpy of bulk diffusion is $\Delta H_b = (2.1 \pm 0.2)\,\text{eV}$ and significantly lower compared to other oxide systems of comparable structures. DFT calculations for the individual cation jumps in m-HfO$_2$ give mostly migration enthalpies of $3\,\text{eV}$ to $5\,\text{eV}$, which agrees with the values obtained for other AO_2 systems. However, two jumps have significantly lower values ($\approx 2\,\text{eV}$) and allow long-range diffusion through the bulk. It is argued that the other jumps occur far less frequently and the DFT results thus agree with the experiments. The difference in activation enthalpy of bulk diffusion between other AO_2 systems and m-HfO$_2$ is attributed to the structural perturbations in the monoclinic system, which are hypothesised to increase ion mobility for immobile ions (such as cations in oxide-ion conducting AO_2 systems). MD simulations in c-HfO$_2$ are able to reproduce the activation enthalpy of bulk diffusion determined experimentally and with DFT. However, molecular static simulations instead produce results much closer to those of other cubic oxide systems. A cooperative migration mechanism of oxygen and hafnium vacancies is proposed and some support is produced, but further study is required.

The observed grain-boundary diffusion activation enthalpy is the same as the activation enthalpy for bulk diffusion with $\Delta H_{gb} = (2.1 \pm 0.3)\,\text{eV}$. This contradicts the traditional picture of fast grain-boundary diffusion along the grain-boundary core and instead fast cation diffusion along space-charge layers is suggested. This theory is supported by the prediction of a reasonable space-charge potential for the investigated system.

Chapter 6

Electrical Properties and Defect Structure of HfO$_2$

Parts of this chapter have been prepared for publication:

M. P. Mueller, F. Gunkel, S. Hoffmann-Eifert, and R. A. De Souza, *in preparation*.

6.1 Introduction

While a profound understanding of the electrical conductivity behaviour is paramount to the search of new functional materials, previous studies focusing on the conductivity of HfO$_2$ fell short when it came to the conductivity in reducing conditions and were unable to come to a uniform assessment of the behaviour in oxidising conditions. For metal–insulator–metal structures, amorphous HfO$_2$ in the insulator layer usually assumes a reduced form due to an applied current, making the understanding of the electrical conductivity behaviour in reducing conditions all the more important. Ellingham diagrams display the thermodynamical stability of compounds relative to their metal. The equilibrium pO_2 at which both HfO$_2$ and pure hafnium metal are stable is $\approx 10^{-100}$ bar at temperatures relevant for the resistive switching process, signifying the importance of the electrical conductivity behaviour in reducing conditions. To date, several authors have looked into the behaviour of hafnias electrical conductance.[145–149] Almost all of them report non-standard behaviour, which differs from the traditional behaviour described in more detail below.

Kofstad and Ruzicka[145] propose m-HfO$_2$ to be an ionic conductor and that oxygen vacancy transport and oxygen interstitial transport are coupled, which is faster than an independent transport. They also report an unusual and complex pressure dependence. However, they reported a very low density of the samples investigated, and porous samples might introduce a large additional surface, affecting the conductivity. Furthermore, they do not extract the true bulk conductivity from their measured values. Tallan et al.[146] found electronic conductivity above oxygen partial pressures of 10^{-6} bar and claim that completely ionized hafnium vacancies and electron holes are the predominant defect species. At lower partial pressures a pO_2-independent source of electronic charge carriers allegedly dominates and results in a broad shallow minimum. They did not measure the same oxygen partial pressure range for all temperatures. Guillot and Anthony[147] measured ionic transport numbers in a comparatively small range of $-10 \leq \log(pO_2/\mathrm{atm}) \leq 0$ and find electronic and ionic contributions to the conductivity. They claim that Schottky disorder agrees with their results. Kharton et al.[148] found predominantly oxygen ionic conductivity, with p-type (electron holes) electronic conductivity increasing with a characteristic slope of $+1/4$. All previous studies neglect the very low oxygen partial pressure regime, which is relevant to the understanding of memristors. Ko et al.[149] investigated nanoscale hafnia films up to an oxygen partial pressures of 10^{-18} bar and found large ionic contributions. They state the importance of charged point defects other than oxygen vacancies, such as oxygen interstitials or hafnium vacancies, due to the weak dependency of conductance on oxygen partial pressure. Electronic conductance is attributed to p-type conduction with a characteristic slope of $+1/11$ to $+1/14$.

Computational studies on trapping in hafnia are available and they predict trapping reactions and defect energy levels of interstitial and vacancy defects.[86,150–154] Additionally, experimental investigations into the defect levels of oxygen vacancies in HfO$_2$ are consistent with computational data.[155–158] However, these studies are not directly applicable to the studies on the conductivity behaviour in m-HfO$_2$. Instead, they may support defect-chemical calculations to describe the defect model in m-HfO$_2$. Evidently, a complete understanding of the defect structure and electrical conductivity behaviour of HfO$_2$ has not yet been achieved. This work aims to contribute to a better understanding of these prop-

6.1 Introduction

erties. To that end, the conductivity was measured as a function of oxygen partial pressure in dense m-HfO$_2$ ceramics at temperatures in the range of $1050 \leq T/\text{K} \leq 1200$ and oxygen partial pressures in the range of $-20 \leq \log(pO_2/\text{bar}) \leq -2$. Numerical defect-chemical calculations are employed to try and understand the observed conductivity behaviour.

6.1.1 Acceptor-doped AO_2 Oxides

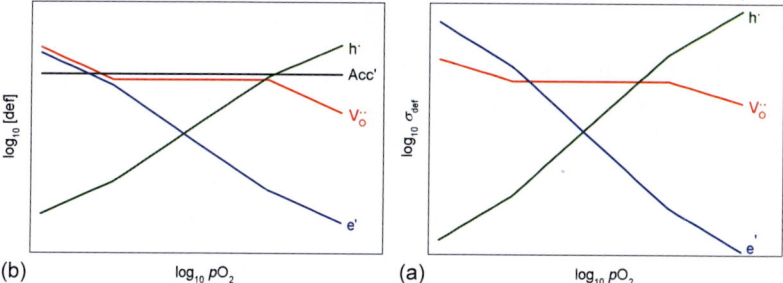

Figure 6.1: Schematic diagram of (a) defect concentrations; (b) partial ionic and electronic conductivities in AO_2 as a function of pO_2.

The commonly expected behaviour of the defect concentrations as a function of pO_2 in acceptor-doped AO_2 oxides is shown in Fig. 6.1(a). The analogous behaviour of the conductivities is shown in Fig. 6.1(b). The conductivity dependence from oxygen partial pressure pO_2 (in place of oxygen activity) is connected to the equilibrium constant of reduction K_{red}:

$$K_{\text{red}} = \frac{[V_O^{\bullet\bullet}][e']^2 pO_2^{1/2}}{[O_O^\times]}. \tag{6.1}$$

The square brackets denote site fractions instead of concentrations here. Under strongly reducing conditions, electrons and oxygen vacancies are the majority defects and the respective law of mass action describing the concentration of the majority defect species is

$$[e'] = \left(\frac{2[O_O^\times]K_{\text{red}}}{pO_2^{1/2}}\right)^{1/3} \propto pO_2^{-1/6}. \qquad (6.2)$$

The concentrations of electrons and oxygen vacancies varies with a characteristic slope of $-\frac{1}{6}$ in the double logarithmic plot due to a deficiency of oxygen. The hole concentration behaves in an opposite way to that of the electron concentration due to the electronic excitation relation:

$$K_e = [h^\bullet][e']. \qquad (6.3)$$

The conductivity in reducing conditions is dominated by electronic conductivity, because electronic species are usually more conductive than ionic species. At higher partial pressures, anti-Frenkel disorder dominates and acceptors and oxygen vacancies become the majority defects. Subsequently, Eq. 6.2 changes to:

$$[e'] = \left(\frac{[O_O^\times]K_{\text{red}}}{[V_O^{\bullet\bullet}]pO_2^{1/2}}\right)^{1/2} \propto pO_2^{-1/4}. \qquad (6.4)$$

The characteristic slope of the hole and electron concentrations increases to $\pm\frac{1}{4}$, while the concentration of oxygen vacancies stays fixed due to the fixed concentration of acceptors. Because holes and electrons are minority defects in this pressure region, the partial ionic conductivity dominates the total conductivity.

The concentration of holes continues to increase with pO_2 and the electronic conductivity begins to overtake the ionic conductivity again at even higher oxygen partial pressures, subsequently dominating the total conductivity:

$$[h^\bullet] = \left(\frac{K_e pO_2^{1/2}}{2[O_O^\times]K_{\text{red}}}\right)^{1/3} \propto pO_2^{1/6}, \qquad (6.5)$$

The concentration of electron holes increases with a characteristic slope of $+\frac{1}{6}$, while the concentration of oxygen vacancies decreases with a characteristic slope of $-\frac{1}{6}$ due to the oxygen excess.

6.2 Experimental Details

Ceramic samples of m-HfO$_2$ were prepared by sintering ball milled, commercial powder (99.95 % purity excluding zirconium, Alfa Aesar®) at $T = 1823\,\mathrm{K}$ for 72 h. The density of the samples was at least 96 % of the theoretical density, measured by Archimedes' principle. X-ray diffractograms measured with a Theta–Theta diffractometer (STOE & Cie GmbH, Darmstadt, Germany) indicated single-phase m-HfO$_2$. For the conductivity experiments the High Temperature Equilibrium Conductance (HTEC) was measured using a custom four-point measurement setup based on an yttria-stabilized ZrO$_2$-oxygen pump.[58,59] A 5 mm × 10 mm-wide and 500 µm thin m-HfO$_2$ sample was connected to the setup by sputtering four platinum contacts on top of the sample. The sputtered contacts were 800 µm wide and ranged over the entire width of the sample. The inner electrodes were spaced 5 mm apart, leading to a square sample geometry of 5 mm × 5 mm for the measured voltage drop. Then, four thin slits were cut into the sample along the contacts and 100 µm thin platinum wire was wrapped around the slits. Platinum paste was applied to the wire and treated at $1243\,\mathrm{K}$ to improve contact. HTEC measurements were done at temperatures in the range of $1050 \leq T/\mathrm{K} \leq 1200$ in an Al$_2$O$_3$ tube furnace and in an oxygen partial pressure range of $-20 \leq \log(p\mathrm{O}_2/\mathrm{bar}) \leq -2$. HTEC measurements give the conductance G instead of the conductivity σ. To obtain the conductivity, the length $l = 5\,\mathrm{mm}$, width $w = 5\,\mathrm{mm}$ and thickness $h = 500\,\mathrm{µm}$ of the sample geometry are needed:

$$\sigma = G \frac{l}{wh}. \tag{6.6}$$

6.3 Results

Fig. 6.2 shows the obtained conductivity σ for all four temperatures $1050 \leq T/\mathrm{K} \leq 1200$ versus the oxygen partial pressure $p\mathrm{O}_2$ as open squares in a double logarithmic plot. In reducing conditions, below $10^{-16}\,\mathrm{bar}$, the conductivity increases with pressure and, disregarding the two deviating points for $1200\,\mathrm{K}$, this behaviour seems to be largely independent from temperature, since the data points converge at the lowest pressures. The two data points in question for

1200 K exhibited connectivity issues and can therefore be neglected. The behaviour in the reducing regime starkly contrasts the previously described expected behaviour for AO_2 oxides. Generally, the conductivity is expected to decrease with a characteristic slope of $-1/4$ or $-1/6$. This behaviour can not be observed in Fig. 6.2 and instead all isotherms see an increase in conductivity with a slope of *ca.* $+1/13$ (for 1150 K) is seen. Although it seems as if all temperatures show the same behaviour, the characteristic slopes in this regime vary, most likely due to statistical errors.

At oxidising conditions, the conductivity increases. This increase starts at 10^{-7} bar for 1050 K and shifts to later pressure values for the higher temperatures. A measurement at 10^{-2} bar is only available for 1150 K, as the other samples could not be measured at this pressure. The behaviour in the oxidising regime resembles the previously expected behaviour for AO_2 oxides. Taking the points at 10^{-2} bar

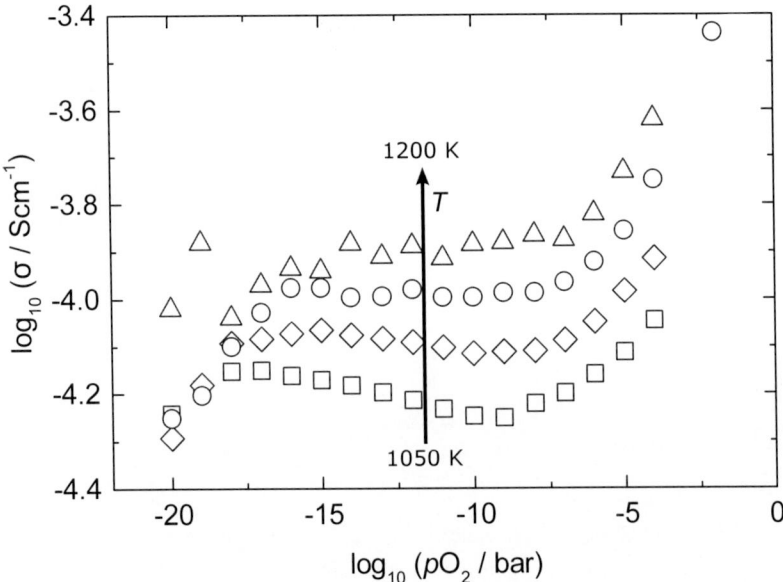

Figure 6.2: Conductivity isotherms plotted against oxygen partial pressure for temperature range of $1050 \leq T/K \leq 1200$ and oxygen partial pressure range of $-20 \leq \log(pO_2/\text{bar}) \leq -2$.

and 10^{-4} bar into account, a characteristic slope of *ca.* $+1/6$ is obtained. This could indicate electronic conductivity by electron holes in this regime. It seems as if all temperatures display a similar increase.

In the central pressure region from 10^{-16} bar to 10^{-7} bar, a plateau in conductivity can be seen, with the extent of the plateau being more pronounced at higher temperatures. This stems from the fact that after the initial increase in conductivity in the reducing regime, the conductivity is found to decrease until it reaches the plateau at $\approx 10^{-11}$ bar. However, this behaviour is only found for $1050\,\text{K}$ and $1100\,\text{K}$ and can not be observed for $1150\,\text{K}$ and $1200\,\text{K}$, where the plateau starts immediately after the initial increase. It is also significantly pronounced for $1050\,\text{K}$, where the plateau is barely visible. Measurements were done in order of decreasing pressure and increasing temperature, so that the sample might have been in a non-equilibrium state in the pressure regime between 10^{-12} bar to 10^{-2} bar, which would result in an erroneously low conductivity. Since the total conductivity is constant over the length of the plateau, it is independent of oxygen partial pressure and is therefore attributed to ionic conductivity. The width of the plateau is also influenced by the beginning of the conductivity increase in the oxidising regime, which starts later for the higher temperatures.

6.4 Discussion

6.4.1 Conductivity Measurements

Fig. 6.3(a) shows the conductivity data measured in this work (A) for comparison, while (b), (c) and (d) show the conductivity data for HfO$_2$ ceramics from Kofstad and Ruzicka[145] (B), Tallan et al.[146] (C) and Guillot and Anthony[147] (D). Note the comparatively small difference in conductivity for the measurements done in this thesis. (D) shows ionic (dashed yellow line) and electronic (solid yellow line) conductivity separately. All three studies investigated higher temperatures than this work and did not reach oxygen partial pressures as low as 10^{-20} bar. They

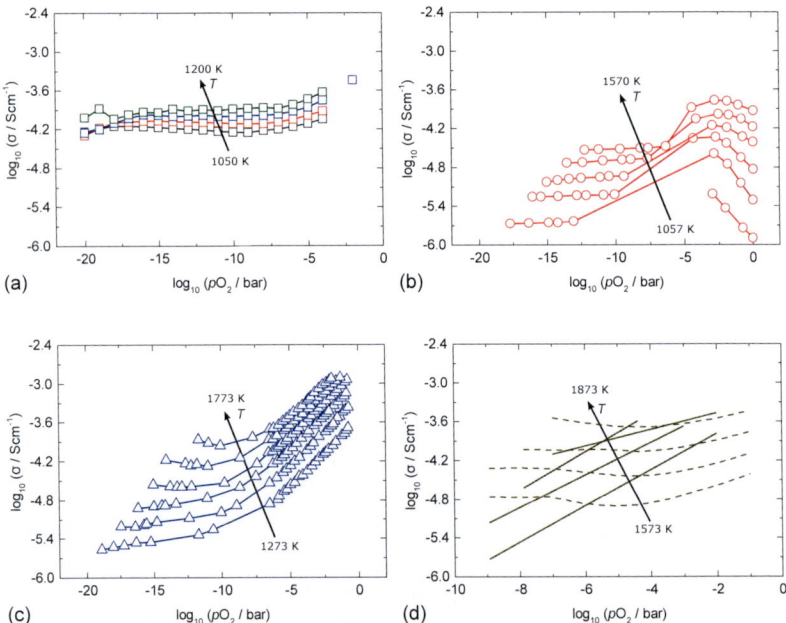

Figure 6.3: (a) Conductivity data found in this study; (b) Conductivity data from Kofstad and Ruzicka[145]; (c) conductivity data from Tallan et al.[146]; (d) conductivity data from Guillot and Anthony[147] (solid line for electronic conductivity, dashed line for ionic conductivity).

6.4 Discussion

found lower total conductivities, but report a similar conductivity curve for the middle and oxidising pressure region. Something resembling a plateau in the pressure regime of *ca.* 10^{-15} bar to 10^{-10} bar and an increase in conductivity when going to higher pressures is found in all studies. (B) finds an additional decrease in conductivity close to standard conditions. The plateau is the least visible in (C), but one isotherm in the middle of the temperature range shows a plateau. While (D) does not look at pressure regimes below 10^{-9} bar, the other two studies do not report the decrease in conductivity in the reducing regime found in this work. The ionic conductivity in oxides is generally attributed to doubly positively charged oxygen vacancies. It is possible to extract an activation energy of diffusion from the ionic conductivity, which can then be compared to other conductivity data and diffusion data from literature. To obtain the activation enthalpy of diffusion, the Nernst-Einstein equation is employed (see also Eq. 2.30 and 2.31). The conductivity-temperature product σT is directly proportional to the ionic mobility μ and the self diffusion coefficient D:

$$D = \frac{D^*}{f^*} = \frac{\mu T k_B}{zq} = \frac{\sigma T k_B}{z^2 q^2 N}, \quad (6.7)$$

with N as the concentration of oxygen vacancies, z as the number of charge carriers, q as the elementary charge, D^* as the tracer diffusion coefficient and f^* as the tracer correlation factor. According to Eq. 6.7 the slope of σT versus the inverse temperature gives the activation enthalpy of diffusion. Fig. 6.4 shows this plot and the resulting activation enthalpy for an oxygen partial pressure of 10^{-10} bar for the conductivity measurements done here and in the literature (A, B and C). Additional values are generated from Molecular Dynamics (MD) simulations[29] (D) and Secondary Ion Mass Spectrometry (SIMS) measurements[81] of oxygen diffusion in m-HfO$_2$ (E).

The slope of the Arrhenius plot results in an activation energy of diffusion of $E_A = (0.70 \pm 0.01)$ eV for the conductivity experiments done in this work. This value is somewhat dependent on the chosen pO_2, as the ionic plateau is barely visible at 1050 K and a range of pO_2 values can refer to the ionic regime. The activation energies obtained from the ionic regimes of the other conductivity studies are significantly higher with $E_A = (1.05 \pm 0.08)$ eV for (B) and $E_A = (1.25 \pm 0.16)$ eV for (C). However, the activation enthalpy obtained here is in good agreement with computational diffusion (D) results,[29] which report $E_A = 0.66$ eV. Exper-

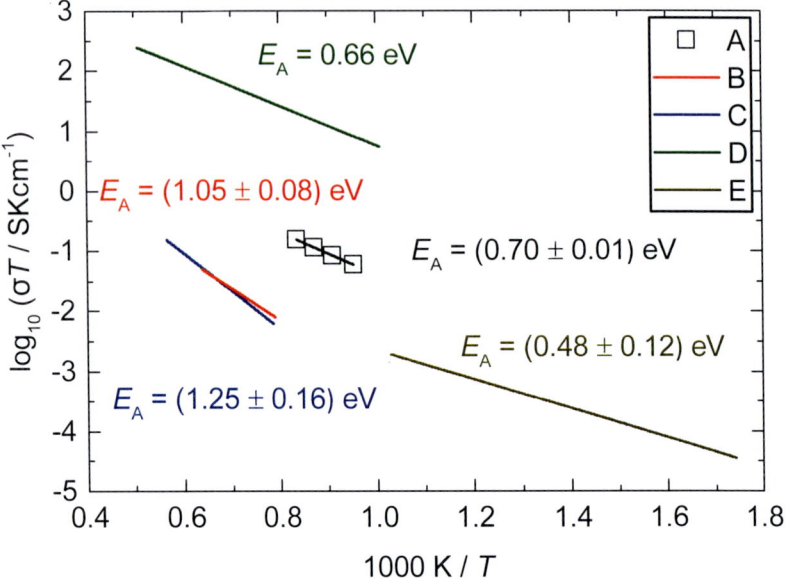

Figure 6.4: σT versus inverse temperature for an oxygen partial pressure of 10^{-10} bar. The activation enthalpy of diffusion is extracted from the slopes of the plot. (A) This study; (B) ref. 145; (C) ref. 146; (D) ref. 29; (E) ref. 81.

imental diffusion results (E)[81] report a values of $E_A = (0.48 \pm 0.12)$ eV, which is smaller than the results obtained from the conductivity experiments (though with a rather significant error), even though they refer to the same batch of HfO_2 ceramics. The reason for this is unclear, but it might be a statistical effect in the conductivity experiments. The sample size is small and the activation energy of diffusion changes significantly in response to slight variances in σT. Furthermore, the experimental results offer a direct look into oxygen diffusion while the conductivity experiments only offer an indirect look and other factors aside from doubly charged oxygen vacancies might influence the conductivity in the ionic regime. The conductivities obtained from MD are significantly higher than the values obtained from other studies. This difference stems most likely from the large amount of oxygen vacancies employed in the MD study (0.625 %) compared to the amount of vacancies in ceramic samples. The experimental diffusion study exhibits lower conductivities, but gives similar results to the two other conductivity studies when extrapolated to higher temperatures. The conductivities deter-

6.4 Discussion

mined here lie between the MD results and the experimental diffusion study. The origin of these differences is not immediately apparent. (B) reports a significantly lower density for their ceramic HfO_2 samples than the literature which might contribute to the difference in activation enthalpy. (C) claims that the overall conductivity is essentially all electronic, with ionic contribution being less than a few percent of the total. Indeed, their results deviate the most from the plateau-type conductivity seen in the other studies. (D) do assume a predominantly ionic conductivity. While they report a plateau in the 10^{-12} bar region, it is unclear why the activation enthalpy of diffusion found is unable to reproduce the other presented studies.

To explain the complex conductivity data determined by the HTEC measurements and understand the defect behaviour responsible, numerical defect-chemical calculations were employed. These give the defect concentrations in the sample and subsequently the conductivity of the different defect species, assuming the defect mobilities are known. All calculations were done for an acceptor dopant level of $5.3 \times 10^{18}\,\text{cm}^{-3}$. This value is obtained by extrapolating a mobility for $V_O^{\bullet\bullet}$ from (D) and comparing it to the measured conductivity values in the ionic regime, therefore approximating a defect concentration. To calculate the total conductivity from the defect concentrations, mobility values found by Sasaki and Maier[159] for the electronic species in yttria-stabilised ZrO_2 were adopted. Mobility values for the doubly positively charged oxygen vacancy species were adopted from diffusivity values from (D).

In order to then describe the observed conductivity behaviour, first the expected behaviour depicted in Fig. 6.1 is emulated. Eq. 6.1 and 6.3 should generally suffice to obtain a reasonable description of this behaviour. As Fig. 6.5(a) shows, this is not the case here.

While the ionic region and the electronic region in the oxidising regime are reproduced relatively well, the reducing regime and its decrease in conductivity can not be reproduced by this model. A reduction enthalpy of $K_\text{red} = 7.25\,\text{eV}$ has been used for the shown isotherms, which is somewhat higher than the enthalpies found in similar oxides.[116] Reducing the reduction enthalpy leads to a behaviour according to Fig. 6.1(b), but not the behaviour found here. A band gap of $4.0\,\text{eV}$ was assumed for the calculations, as it reproduces the p-type conductivity in the oxidising regime much better than the experimentally reported band

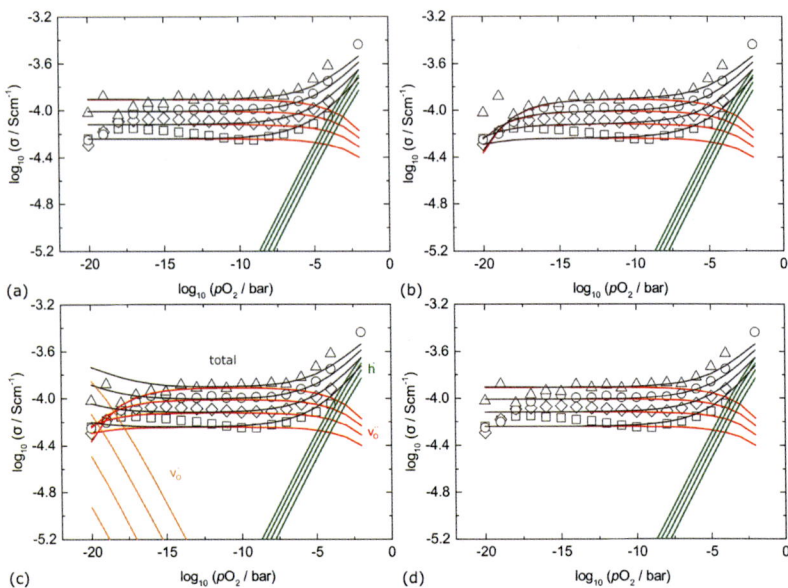

Figure 6.5: Conductivity contributions plotted against oxygen partial pressure obtained from defect-chemical calculations for temperature range of $1050 \leq T/\mathrm{K} \leq 1200$ and oxygen partial pressure range of $-20 \leq \log(p\mathrm{O}_2/\mathrm{bar}) \leq -2$. The different graphs are obtained from different calculations. Red lines refer to the conductivity associated with the $V_\mathrm{O}^{\bullet\bullet}$ species; green lines to $h^\bullet$; orange lines to $V_\mathrm{O}^{\bullet}$ and grey lines to the total conductivity. All calculations contain the electronic excitation equation and the reduction equation; (a) no additional equation; (b) ionisation of $V_\mathrm{O}^{\bullet}$ to $V_\mathrm{O}^{\bullet\bullet}$; (c) the same as (b) but the $V_\mathrm{O}^{\bullet}$ species is considered to be a charge carrier; (d) ionisation of $V_\mathrm{O}^{\times}$ to $V_\mathrm{O}^{\bullet\bullet}$.

gap of $5.7\,\mathrm{eV}$ for m-HfO$_2$.[153,156] This difference is most likely connected to the mobility of holes taken from ref. 159, which investigated yttria-stabilised ZrO$_2$ and not m-HfO$_2$. A hole mobility determined specifically for m-HfO$_2$ could also lead to a better description of K_e. It is not possible to reproduce the reducing regime found here with only Eq. 6.1 and 6.3. Thus, another defect reaction is needed.

When doubly positively charged oxygen vacancies take up electrons in the reducing regime, the concentration of doubly positively charged vacancies is decreased in favour of the concentration of singly positively charged vacancies.

6.4 Discussion

Subsequently, the conductivity begins to be primarily ionic where normally an electronic conductivity would be expected. The generation of singly positively charged oxygen vacancies is described by an ionisation equation:

$$K_{V_O^{\bullet}/V_O^{\bullet\bullet}} = \frac{[V_O^{\bullet\bullet}][e']}{[V_O^{\bullet}]}. \tag{6.8}$$

Fig. 6.5(b) shows the results of these calculations with $K_{V_O^{\bullet}/V_O^{\bullet\bullet}} = 2.0\,\text{eV}$. This addition describes the conductivity behaviour in the reducing regime significantly better. The singly positively charged oxygen vacancies are not contributing to the total conductivity here. If they play an important role as charge carriers, their mobility has to be determined. Duncan et al.[160] claim that the fully ionised oxygen vacancies in m-HfO$_2$ exhibit a significantly lower migration barrier compared to the singly positively charged oxygen vacancies. One possible deduction then is, that the conductivity of $V_O^{\bullet\bullet}$ is higher than that of $V_O^{\bullet}$. However, Nakayama et al.[161] find that doubly positively charged oxygen vacancies attract localised electrons and form associates, which can be interpreted as a singly positively charged oxygen vacancy. The migration energy of the vacancies in these associates is substantially lowered, which indicates that singly positively charged oxygen vacancies might be more mobile compared to doubly charged vacancies, if they do contribute to the ionic conductivity in a meaningful way.

Fig. 6.5(c) shows the results for this assumption, which results in a worse description of the experimental data compared to Fig. 6.5(b). Subsequently, singly positively charged oxygen vacancies are most likely not charge carriers themselves and their generation only serves to reduce the amount of doubly charged vacancies regarding the total conductivity.

The question arises whether singly positively charged oxygen vacancies are responsible for this behaviour or if neutral oxygen vacancies are the reason. To answer this, neutral oxygen vacancies are implemented instead of singly charged vacancies according to the following (analogous) equation:

$$K_{V_O^{x}/V_O^{\bullet\bullet}} = \frac{[V_O^{\bullet\bullet}][e']^2}{[V_O^{x}]}. \tag{6.9}$$

This defect reaction does not change the total conductivity behaviour visibly compared to the most simple case (Fig. 6.5(a)), as can be seen in Fig. 6.5(d). While neutral oxygen vacancies are obviously no charge carriers, the implementation according to Eq. 6.9 also does not change the concentrations of the other defect species significantly unless inflated values of $K_{V_O^x/V_O^{\bullet\bullet}}$ are chosen. Even though the concentration of neutral oxygen vacancies at 10^{-20} bar is non-trivial with $[V_O^x] \approx 10^{18}\,\mathrm{cm}^{-3}$, the concentration of other dominant defect species is not affected due to the total amount of oxygen atoms being orders of magnitude higher. Literature additionally reports on the existence of singly and doubly negatively charged oxygen vacancies.[150–153] These are not taken into account here, being unlikely to form under the conditions studied in this work.

The introduction of anti-Frenkel defects was attempted as well, but the enthalpy of formation is relatively high with $E_{\mathrm{a-Fr}} = 5.47\,\mathrm{eV}$.[86] Other publications point to similar values.[141,142] Subsequently, the total conductivity is not changed by this defect reaction, since the amount of oxygen interstitials is in the order of magnitude of $10^4\,\mathrm{cm}^{-3}$ and therefore too small to have significant impact. The formation of Frenkel and Schottky defects is neglected as well, since their formation enthalpies are even larger.[86] The amount of hafnium vacancies at the comparatively low temperatures investigated here is negligible at a Schottky enthalpy of $E_{\mathrm{Sch}} = 7.13\,\mathrm{eV}$.

Fig. 6.6 shows the calculated total conductivities from Fig. 6.5(b) compared to the experimental results. The model presented here is able to reproduce the conductivity at high temperatures very well. The inflection of the conductivity seen at 10^{-16} bar at lower temperatures is neglected and not reproducible by the chosen model since its origin is not known. The electronic conductivity in oxidising conditions is sufficiently well reproduced.

Table 6.1 displays the defect reaction energies, pre-exponential parameters and mobilities used for the defect-chemical calculations.

6.4 Discussion

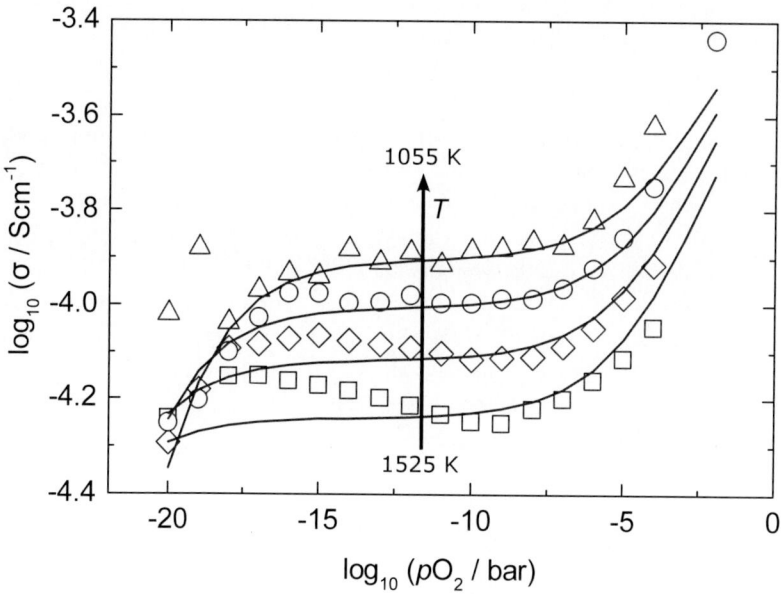

Figure 6.6: Conductivity isotherms plotted against oxygen partial pressure for temperature range of $1050 \leq T/K \leq 1200$ and oxygen partial pressure range of $-20 \leq \log(pO_2/\mathrm{bar}) \leq -2$. Squares are the measured values, lines are obtained from defect-chemical calculations.

Table 6.1: Defect chemical parameters in m-HfO$_2$, with $k = 8.617 \times 10^{-5}\,\mathrm{eV/K}$.

Charge carrier	Expression
Oxygen vacancy	$K_{\mathrm{red}} = 8.35 \times 10^{56} \cdot \exp(-7.25\,\mathrm{eV}/kT)$
	$\mu_{V_O^{\bullet\bullet}} = 100.029/T \cdot \exp(-0.657\,\mathrm{eV}/kT)$
Electron	$K_{V_O^{\bullet}/V_O^{\bullet\bullet}} = 2.89 \times 10^{28} \cdot \exp(-2.0\,\mathrm{eV}/kT)$
	$K_{V_O^{\times}/V_O^{\bullet\bullet}} = 2.89 \times 10^{28} \cdot \exp(-2.0\,\mathrm{eV}/kT)$
	$\mu_e = 6300/T \cdot \exp(-0.55\,\mathrm{eV}/kT)$
Hole	$\mu_h = 1.8/T \cdot \exp(-0.1\,\mathrm{eV}/kT)$
Electron + hole	$K_e = 1.67 \times 10^{57} \cdot \exp(-4.0\,\mathrm{eV}/kT)$

6.5 Summary

The defect structure of monoclinic HfO_2 (m-HfO_2) was studied by means of oxygen partial pressure dependent HTEC measurements. In reducing conditions the total conductivity was found to increase with oxygen partial pressure instead of decreasing, as would be expected for similar fluorite-type oxides. Numerical defect-chemical calculations show that singly positively charged oxygen vacancies are most likely responsible for this behaviour. In middling oxygen partial pressure regimes ionic conductivity is dominant and the activation enthalpy of diffusion is calculated to be $E_A \approx 0.7\,\text{eV}$. This agrees well with experimental and computational results for oxygen diffusion in m-HfO_2. In oxidising conditions the total conductivity increases with oxygen partial pressure due to electron holes. An inflection of the conductivity is found at low oxygen partial pressures and temperatures, but its origin could not be fully explained.

Chapter 7

Conclusion

The aim of this thesis was to further the neglected understanding of ion transport in monoclinic HfO_2 (m-HfO_2) in order to better understand and improve future electronic devices. To that end, oxygen and cation diffusion were directly investigated experimentally and cation diffusion in m-HfO_2 and cubic HfO_2 (c-HfO_2) was studied computationally with first-principles calculations, Molecular Dynamics (MD) simulations and molecular static simulations. In addition, High Temperature Equilibrium Conductance (HTEC) measurements were employed together with numerical defect-chemical calculations to understand the conductive behaviour of m-HfO_2.

In the first part of this thesis, oxygen ion diffusion was investigated. Oxygen isotope depth profiles in ceramic samples of m-HfO_2 were directly determined by means of ($^{18}O/^{16}O$) isotope exchange annealing and Secondary Ion Mass Spectrometry (SIMS) measurements for the first time. All isotope profiles were found to exhibit behaviour more complicated than simple bulk diffusion. Two features were assigned to the profiles: the first feature, closer to the surface, displayed significantly slower diffusion than the second feature found deeper in the sample. The profile behaviour could have been attributed to either Harrison-type-B diffusion kinetics or slow diffusion in an equilibrium space-charge layer, but both of these explanations were shown to be unlikely. Instead, the first feature was attributed mainly to slow oxygen diffusion in an impurity silicate phase. The second feature was attributed to oxygen diffusion in bulk m-HfO_2. Finite-Element-Method (FEM) simulations were utilised to describe the entirety of the isotope profiles and obtain diffusion coefficients. The activation enthalpy of oxygen tracer diffusion in bulk m-HfO$_2$ was found to be $\Delta H_{D^*} \approx 0.5\,\text{eV}$, which is in good agreement with computational simulations. This shows that any applications of HfO$_2$ need to be wary of inevitable silicon impurities in order to avoid undesirable effects on the oxygen diffusion in HfO$_2$.

The second part aimed to investigate the diffusion of cations in m-HfO$_2$. A low-temperature preparation method, atomic layer deposition, was used to prepare non-equilibrium samples. Diffusion annealing with zirconium as the tracer were then performed on the thin film samples and isotope diffusion profiles were obtained by subsequent SIMS measurements. These profiles displayed two features, which were analysed by FEM simulations under the assumption that fast grain-boundary diffusion took place. The obtained activation enthalpy of bulk diffusion was found to be $\Delta H_b \approx 2.1\,\text{eV}$, significantly lower compared to other AO_2 systems. The difference in activation enthalpy was attributed to the structural perturbations in the monoclinic system, which are hypothesised to increase ion mobility for immobile ions (such as cations in oxide-ion conducting AO_2 systems). The activation enthalpy of grain-boundary diffusion obtained from the profiles was the same as the activation enthalpy for bulk diffusion with $\Delta H_{gb} \approx 2.1\,\text{eV}$. This contradicts the traditional picture of fast grain-boundary diffusion along the grain-boundary core, and fast diffusion along space-charge layers was suggested instead, supported by a reasonable space-charge potential calculated for the investigated system. To take a deeper look into the low activation enthalpy of cation diffusion various computational approaches have been undertaken. The migration barriers of individual cation jumps in m-HfO$_2$ were investigated by Density-Functional-Theory (DFT) calculations. Two jumps have been shown to exhibit similar values as the experiments ($\approx 2\,\text{eV}$) while being capable of migrating through the bulk. The other jumps are expected to occur far less frequently and the DFT results thus agree with the experiments. MD simulations were employed to study hafnium diffusion in c-HfO$_2$ because m-HfO$_2$ was not stable during the simulations. No diffusion was found to take place in the investigated temperature range at first, due to the generally low diffusivity of cations. By applying an electrical field to the system, diffusion could be enhanced and mobility values for hafnium diffusion in c-HfO$_2$ were obtained. Extrapolation of the mobility to a field strength of zero yielded an enthalpy of migration of $\Delta H_{mig} \approx 2.2\,\text{eV}$, which agrees well with experimental and DFT results. Molecular static simulations painted a different picture and resulted in a migration barrier of $\Delta H_{\text{mig}} \approx 6.6\,\text{eV}$, much closer to other cubic AO_2 systems. Oxygen vacancy formation has been observed during all simulations and static lattice simulations have shown that hafnium and oxygen vacancies interact. A coupled mechanism of hafnium vacancies and oxygen vacancies is therefore investigated as a possible explanation for the difference of molecular static and dynamic simulations.

Finally, the electrical conductivity behaviour of m-HfO$_2$ was studied in the third part of this thesis. In reducing conditions the total conductivity was found to increase with oxygen partial pressure, a rather surprising behaviour. It was attributed to an accumulation of singly positive charged oxygen vacancies, that do not contribute to the total conductivity. In middling oxygen partial pressure regimes ionic conductivity was found to be dominant and the activation enthalpy of diffusion was calculated to be $E_A \approx 0.7\,\text{eV}$. This agrees well with known values for the activation enthalpy of oxygen diffusion in m-HfO$_2$. In oxidising conditions the total conductivity increases with oxygen partial pressure due to the formation of electron holes, which is the expected behaviour for AO_2 oxides. Not all aspects of the electrical conductivity behaviour of m-HfO$_2$ have been understood so far and additional experiments with more temperatures and higher and lower partial pressures are a strong point to improve upon.

As a result of this thesis, much insight has been won into the diffusion of both anions and cations in HfO$_2$. The diffusion of oxygen and zirconium in m-HfO$_2$ was directly investigated for the first time and complicated isotope profiles were described by numerical models. A generally neglected alternative to the traditional explanation of fast grain-boundary diffusion, fast diffusion along space-charge layers, is discussed and supporting evidence is presented. The importance of oxygen vacancies other than the doubly positively charged vacancy for the conductivity behaviour of m-HfO$_2$ has also been discussed. Furthermore, two new methods for investigating minority defect diffusion in oxides have been proposed, one experimental and one computational. The use of non-equilibrium samples in diffusion experiments allows easier investigation of minority defect diffusion and the application of field-enhanced ion transport allows determination of migration enthalpies of minority defects without increasing the simulation temperatures to unreasonable levels. Future work will undoubtedly be able to expand upon the results presented here and make use of the models and methods described.

Bibliography

[1] V. A. Gritsenko, T. V. Perevalov, and D. R. Islamov, Phys. Rep. **613**, 1 (2016).

[2] E. P. Gusev, D. A. Buchanan, E. Cartier, A. Kumar, D. DiMaria, S. Guha, A. Callegari, S. Zafar, P. C. Jamison, and D. A. Neumayer, IEDM - Tech. Dig. (2001).

[3] T. S. Böscke, J. Müller, D. Bräuhaus, U. Schröder, and U. Böttger, Appl. Phys. Lett. **99**, 102903 (2011).

[4] G. D. Wilk, R. M. Wallace, and J. M. Anthony, J. Appl. Phys. **89**, 5243 (2001).

[5] J. Robertson, Appl. Surf. Sci. **190**, 2 (2002).

[6] J. Robertson, O. Sharia, and A. A. Demkov, Appl. Phys. Lett. **91**, 132912 (2007).

[7] W. Li, J. Zhou, S. Cai, Z. Yu, J. Zhang, N. Fang, T. Li, Y. Wu, T. Chen, X. Xie, H. Ma, K. Yan, N. Dai, X. Wu, H. Zhao, Z. Wang, D. He, L. Pan, Y. Shi, P. Wang, W. Chen, K. Nagashio, X. Duan, and X. Wang, Nat. Electron. **2**, 563 (2019).

[8] R. Waser, R. Dittmann, G. Staikov, and K. Szot, Adv. Mater. **21**, 2632 (2009).

[9] M. Y. Chan, T. Zhang, V. Ho, and P. S. Lee, Microelectron. Eng. **85**, 2420 (2008).

[10] V. Milo, C. Zambelli, P. Olivo, E. Pérez, M. K. Mahadevaiah, O. G. Ossorio, C. Wenger, and D. Ielmini, APL Mater. **7**, 081120 (2019).

[11] F. Cüppers, S. Menzel, C. Bengel, A. Hardtdegen, M. von Witzleben, U. Böttger, R. Waser, and S. Hoffmann-Eifert, APL Mater. **7**, 091105 (2019).

[12] G. H. Kim, H. Ju, M. K. Yang, D. K. Lee, J. W. Choi, J. H. Jang, S. G. Lee, I. S. Cha, B. K. Park, J. H. Han, T.-M. Chung, K. M. Kim, C. S. Hwang, and Y. K. Lee, Small **13**, 1701781 (2017).

[13] S. Clima, Y. Y. Chen, C. Y. Chen, L. Goux, B. Govoreanu, R. Degraeve, A. Fantini, M. Jurczak, and G. Pourtois, J. Appl. Phys. **119**, 225107 (2016).

[14] P. Calka, M. Sowinska, T. Bertaud, D. Walczyk, J. Dabrowski, P. Zaumseil, C. Walczyk, A. Gloskovskii, X. Cartoixà, J. Suñé, and T. Schroeder, ACS Appl. Mater. Interfaces **6**, 5056 (2014).

[15] M. Lanza, K. Zhang, M. Porti, M. Nafría, Z. Y. Shen, L. F. Liu, J. F. Kang, D. Gilmer, and G. Bersuker, Appl. Phys. Lett. **100**, 123508 (2012).

[16] B. Govoreanu, G. S. Kar, Y.-Y. Chen, V. Paraschiv, S. Kubicek, A. Fantini, I. P. Radu, L. Goux, S. Clima, R. Degraeve, N. Jossart, O. Richard, T. Vandeweyer, K. Seo, P. Hendrickx, G. Pourtois, H. Bender, L. Altimime, D. J. Wouters, J. A. Kittl, and M. Jurczak, in *IEEE International Electron Devices Meeting (IEDM), 2011* (IEEE, Piscataway, NJ, 2011) pp. 31.6.1–31.6.4.

[17] S. Clima, Y. Y. Chen, R. Degraeve, M. Mees, K. Sankaran, B. Govoreanu, M. Jurczak, S. de Gendt, and G. Pourtois, Appl. Phys. Lett. **100**, 133102 (2012).

[18] L. Goux, P. Czarnecki, Y. Y. Chen, L. Pantisano, X. Wang, R. Degraeve, B. Govoreanu, M. Jurczak, D. J. Wouters, and L. Altimime, Appl. Phys. Lett. **97**, 243509 (2010).

[19] O. Ohtaka, H. Fukui, T. Kunisada, T. Fujisawa, K. Funakoshi, W. Utsumi, T. Irifune, K. Kuroda, and T. Kikegawa, J. Am. Ceram. Soc. **84**, 1369 (2001).

[20] Y. Wang, F. Zahid, J. Wang, and H. Guo, Phys. Rev. B **85**, 224110 (2012).

[21] J. Robertson, Eur. Phys. J. Appl. Phys. **28**, 265 (2004).

[22] F. A. Kröger and H. J. Vink, *Relations between the concentrations of imperfections in crystalline solids*, Vol. 3 (Elsevier, 1956).

[23] M. Kilo, M. A. Taylor, C. Argirusis, G. Borchardt, R. A. Jackson, M. Martin, and M. Weller, Solid State Ionics **175**, 823 (2004).

[24] J. Crank, *The Mathematics of Diffusion* (Oxford University Press, 1975).

[25] M. Kilo, C. Argirusis, G. Borchardt, and R. A. Jackson, Phys. Chem. Chem. Phys. **5**, 2219 (2003).

[26] S. Beschnitt, T. Zacherle, and R. A. De Souza, J. Phys. Chem. C **119**, 27307 (2015).

[27] R. A. Jackson, A. D. Murray, J. H. Harding, and C. R. A. Catlow, Phil. Mag. A **53**, 27 (1986).

[28] B. Dorado, D. A. Andersson, C. R. Stanek, M. Bertolus, B. P. Uberuaga, G. Martin, M. Freyss, and P. Garcia, Phys. Rev. B **86**, 035110 (2012).

[29] M. Schie, M. P. Mueller, M. Salinga, R. Waser, and R. A. De Souza, J. Chem. Phys. **146**, 094508 (2017).

[30] E. J. Verwey, Physica **2**, 1059 (1935).

[31] N. F. Mott and R. W. Gurney, *Electronic processes in ionic crystals* (Clarendon Press, 1940).

[32] A. R. Genreith-Schriever and R. A. De Souza, Phys. Rev. B **94**, 224304 (2016).

[33] C. Herzig and Y. Mishin, in *Diffusion in Condensed Matter*, edited by P. Heitjans and J. Kärger (Springer-Verlag Berlin Heidelberg, Berlin, Heidelberg, 2005) pp. 337–366.

[34] D. L. Beke and G. Erdelyi, in *Numerical data and functional relationships in science and technology*, Landolt-Börnstein - Group III Condensed Matter, Vol. 33A, edited by O. Madelung and H. Landolt (Springer, Berlin, 1998) pp. 1–26.

[35] R. W. Balluffi, Phys. Status Solidi (B) **42**, 11 (1970).

[36] H. Mehrer, *Diffusion in Solids: Fundamentals, Methods, Materials, Diffusion-Controlled Processes*, Springer Series in Solid-State Sciences, Vol. 155 (Springer-Verlag GmbH, Berlin Heidelberg, 2007).

[37] I. Kaur, Y. Mishin, and W. Gust, *Fundamentals of grain and interphase boundary diffusion*, 3rd ed. (Wiley, Chichester, 1995).

[38] S. Beschnitt and R. A. De Souza, Solid State Ionics **305**, 23 (2017).

[39] J. P. Parras and R. A. De Souza, Acta Mater. **195**, 383 (2020).

[40] X. Tong, D. S. Mebane, and R. A. De Souza, J. Am. Ceram. Soc. **103**, 5 (2020).

[41] D. van Laethem, J. Deconinck, and A. Hubin, J. Eur. Ceram. Soc. **39**, 432 (2019).

[42] H. J. Avila-Paredes, K. Choi, C.-T. Chen, and S. Kim, J. Mater. Chem. **19**, 4837 (2009).

[43] R. A. De Souza, M. J. Pietrowski, U. Anselmi-Tamburini, S. Kim, Z. A. Munir, and M. Martin, Phys. Chem. Chem. Phys. **10**, 2067 (2008).

[44] X. Guo, W. Sigle, J. Fleig, and J. Maier, Solid State Ionics **154-155**, 555 (2002).

[45] X. Guo and R. Waser, Prog. Mater. Sci. **51**, 151 (2006).

[46] L. G. Harrison, Trans. Faraday Soc. **57**, 1191 (1961).

[47] Y.-C. Chung and B. J. Wuensch, J. Appl. Phys. **79**, 8323 (1996).

[48] Y. Mishin and C. Herzig, Mater. Sci. Eng. A **260**, 55 (1999).

[49] R. A. De Souza and M. Martin, Phys. Chem. Chem. Phys. **10**, 2356 (2008).

[50] R. A. De Souza, Phys. Chem. Chem. Phys. **11**, 9939 (2009).

[51] R. A. De Souza and M. Martin, Bunsen-Magazin **8**, 109 (2006).

[52] R. A. De Souza, J. Zehnpfenning, M. Martin, and J. Maier, Solid State Ionics **176**, 1465 (2005).

[53] P. Fielitz and G. Borchardt, Solid State Ionics **144**, 71 (2001).

[54] M. Kessel, R. A. De Souza, and M. Martin, Phys. Chem. Chem. Phys. **17**, 12587 (2015).

[55] P. de Groot, Adv. Opt. Photon. **7**, 1 (2015).

[56] L. Spieß, R. Schwarzer, H. Behnken, and G. Teichert, *Moderne Röntgenbeugung* (B. G. Teubner Verlag, 2005).

[57] K. C. A. Smith and C. W. Oatley, Br. J. Appl. Phys. **6**, 391 (1955).

[58] C. A. Ohly, *Nanocrystalline alkaline earth titanates and their electrical conductivity characteristics under changing oxygen ambients*, Ph.D. thesis, RWTH Aachen University, Germany (2003).

[59] F. Gunkel, *The role of defects at functional interfaces between polar and non-polar perovskite oxides*, Ph.D. thesis, RWTH Aachen University, Germany (2013).

[60] R. Dronskowski, *Computational chemistry of solid state materials: A guide for materials scientists, chemists, physicists and others*, 1st ed. (Wiley-VCH, Weinheim, 2007).

[61] M. Born and R. Oppenheimer, Ann. Phys. **389**, 457 (1927).

[62] P. Hohenberg and W. Kohn, Phys. Rev. **136**, 864 (1964).

[63] W. Kohn and L. J. Sham, Phys. Rev. **140**, A1133 (1965).

[64] J. P. Perdew, K. Burke, and M. Ernzerhof, Phys. Rev. Lett. **78**, 1396 (1997).

[65] F. Bloch, Z. Physik **52**, 555 (1929).

[66] H. J. Monkhorst and J. D. Pack, Phys. Rev. B **13**, 5188 (1976).

[67] P. E. Blöchl, Phys. Rev. B **50**, 17953 (1994).

[68] H. Jónsson, G. Mills, and K. W. Jacobsen, in *Classical and quantum dynamics in condensed phased simulations*, edited by B. J. Berne, D. F. Coker, and G. Ciccotti (World Scientific Pub. Co, 1998) pp. 385–404.

[69] G. Henkelman, B. P. Uberuaga, and H. Jónsson, J. Chem. Phys. **113**, 9901 (2000).

[70] G. Henkelman and H. Jónsson, J. Chem. Phys. **113**, 9978 (2000).

[71] J. Kaub, J. Kler, S. C. Parker, and R. A. De Souza, Phys. Chem. Chem. Phys. **22**, 5413 (2020).

[72] L. Verlet, Phys. Rev. **159**, 98 (1967).

[73] W. C. Swope, H. C. Andersen, P. H. Berens, and K. R. Wilson, J. Chem. Phys. **76**, 637 (1982).

[74] S. Nosé, J. Chem. Phys. **81**, 511 (1984).

[75] Hoover, Phys. Rev. A **31**, 1695 (1985).

[76] C. R. A. Catlow, Proc. R. Soc. Lond. A **353**, 533 (1977).

[77] R. A. De Souza, M. S. Islam, and E. Ivers-Tiffée, J. Mater. Chem. **9**, 1621 (1999).

[78] M. S. Islam, J. Mater. Chem. **10**, 1027 (2000).

[79] G. Balducci, M. S. Islam, J. Kašpar, P. Fornasiero, and M. Graziani, Chem. Mater. **12**, 677 (2000).

[80] J. H. Harding, Rep. Prog. Phys. **53**, 1403 (1990).

[81] M. P. Mueller and R. A. De Souza, Appl. Phys. Lett. **112**, 051908 (2018).

[82] S. Zafar, H. Jagannathan, L. F. Edge, and D. Gupta, Appl. Phys. Lett. **98**, 2903 (2011).

[83] J. A. Kilner, S. J. Skinner, and H. H. Brongersma, J. Solid State Electrochem. **15**, 861 (2011).

[84] K. R. Whittle, G. R. Lumpkin, and S. E. Ashbrook, J. Solid State Chem. **179**, 512 (2006).

[85] R. A. De Souza, J. Mater. Chem. A **5**, 20334 (2017).

[86] J. X. Zheng, G. Ceder, T. Maxisch, W. K. Chim, and W. K. Choi, Phys. Rev. B **75**, 104112 (2007).

[87] Y. Guo and J. Robertson, Appl. Phys. Lett. **105**, 223516 (2014).

[88] R. A. De Souza and R. J. Chater, Solid State Ionics **176**, 1915 (2005).

[89] R. A. De Souza, Adv. Funct. Mater. **25**, 6326 (2015).

[90] R. A. De Souza and M. Martin, MRS Bull. **34**, 907 (2009).

[91] H. Schraknepper and R. A. De Souza, J. Appl. Phys. **119**, 064903 (2016).

[92] R. A. De Souza, J. A. Kilner, and J. F. Walker, Mater. Lett. **43**, 43 (2000).

[93] R. A. De Souza, V. Metlenko, D. Park, and Thomas E. Weirich, Phys. Rev. B **85** (2012).

[94] R.-V. Wang and P. C. McIntyre, J. Appl. Phys. **97**, 023508 (2005).

[95] M. de Ridder, A. G. J. Vervoort, R. G. van Welzenis, and H. H. Brongersma, Solid State Ionics **156**, 255 (2003).

[96] K. V. Hansen, K. Norrman, and M. Mogensen, Surf. Interface Anal. **38**, 911 (2006).

[97] A. Bernasik, K. Kowalski, and A. Sadowski, J. Phys. Chem. Solids **63**, 233 (2002).

[98] A. E. Hughes and S. P. S. Badwal, Solid State Ionics **40-41**, 312 (1990).

[99] M. J. Pietrowski, R. A. De Souza, M. Fartmann, R. ter Veen, and M. Martin, Fuel Cells **13**, 673 (2013).

[100] H. Kim and P. C. McIntyre, J. Appl. Phys. **92**, 5094 (2002).

[101] S. Stemmer, Y. Li, B. Foran, P. S. Lysaght, S. K. Streiffer, P. Fuoss, and S. Seifert, Appl. Phys. Lett. **83**, 3141 (2003).

[102] L. V. Goncharova, M. Dalponte, D. G. Starodub, T. Gustafsson, E. Garfunkel, P. S. Lysaght, B. Foran, J. Barnett, and G. Bersuker, Appl. Phys. Lett. **89**, 044108 (2006).

[103] M. A. Lamkin, F. L. Riley, and R. J. Fordham, J. Eur. Ceram. Soc. **10**, 347 (1992).

[104] R. Freer, Contr. Mineral. and Petrol. **76**, 440 (1981).

[105] G. Broglia, G. Ori, L. Larcher, and M. Montorsi, Modell. Simul. Mater. Sci. Eng. **22**, 065006 (2014).

[106] N. Capron, P. Broqvist, and A. Pasquarello, Appl. Phys. Lett. **91**, 2905 (2007).

[107] K. McKenna and A. L. Shluger, Appl. Phys. Lett. **95**, 2111 (2009).

[108] M. P. Mueller, K. Pingen, A. Hardtdegen, S. Aussen, A. Kindsmueller, S. Hoffmann-Eifert, and R. A. De Souza, APL Mater. **8**, 081104 (2020).

[109] K. Ando and Y. Oishi, J. Nucl. Sci. Technol. **20**, 973 (1983).

[110] C. Sari, J. Nucl. Mater. **78**, 425 (1978).

[111] B. W. Busch, W. H. Schulte, E. Garfunkel, T. Gustafsson, W. Qi, R. Nieh, and J. Lee, Phys. Rev. B **62**, R13290 (2000).

[112] H. Inaba and H. Tagawa, Solid State Ionics **83**, 1 (1996).

[113] J. C. Boivin and G. Mairesse, Chem. Mater. **10**, 2870 (1998).

[114] J. A. Kilner, Solid State Ionics **129**, 13 (2000).

[115] A. Orera and P. R. Slater, Chem. Mater. **22**, 675 (2010).

[116] T. Zacherle, A. Schriever, R. A. De Souza, and M. Martin, Phys. Rev. B **87**, 134104 (2013).

[117] H. Matzke, J. Chem. Soc., Faraday Trans. 2 **83**, 1121 (1987).

[118] A. Hardtdegen, H. Zhang, and S. Hoffmann-Eifert, ECS Trans. **75**, 177 (2016).

[119] S. P. Waldow, H. Wardenga, S. Beschnitt, A. Klein, and R. A. De Souza, J. Phys. Chem. C **123**, 6340 (2019).

[120] Z. Shen, S. J. Skinner, and J. A. Kilner, Phys. Chem. Chem. Phys. **21**, 13194 (2019).

[121] J. Druce, T. Ishihara, and J. A. Kilner, Solid State Ionics **262**, 893 (2014).

[122] J. Druce, H. Téllez, T. Ishihara, and J. A. Kilner, Faraday Discuss. **182**, 271 (2015).

[123] H. Yokokawa, N. Sakai, T. Kawada, and M. Dokiya, in *Science and technology of zirconia V*, edited by S. P. S. Badwal and M. J. Bannister (Technomic Publ. Co, Lancaster, Pa., 1993) pp. 59–68.

[124] R. W. G. Wyckoff, *Crystal Structures: Fluorite structure*, 2nd ed., Vol. 1 (Interscience Publishers, New York, 1963).

[125] G. Kresse and J. Furthmüller, Phys. Rev. B **54**, 11169 (1996).

[126] G. Kresse and D. Joubert, Phys. Rev. B **59**, 1758 (1999).

[127] S. Plimpton, J. Comput. Phys. **117**, 1 (1995).

[128] J. D. Gale and A. L. Rohl, Mol. Simul. **29**, 291 (2003).

[129] N. F. Mott and M. J. Littleton, Trans. Faraday Soc. **34**, 485 (1938).

[130] A. Stukowski, Model. Simul. Mater. Sci. Eng. **18**, 025016 (2010).

[131] J. Wang and U. Becker, J. Nucl. Mater. **433**, 424 (2013).

[132] M. Kilo, R. A. Jackson, and G. Borchardt, Philos. Mag. **83**, 3309 (2003).

[133] V. Metlenko, A. H. H. Ramadan, F. Gunkel, H. Du, H. Schraknepper, S. Hoffmann-Eifert, R. Dittmann, R. Waser, and R. A. De Souza, Nanoscale **6**, 12864 (2014).

[134] N. Bonanos, unpublished work. Cited by M. Mogensen, D. Lybye, P. V. Hendriksen, F. W. Poulsen, Solid State Ionics **174**, 279 (2004).

[135] U. Brossmann, G. Knoner, H.-E. Schaefer, and R. Würschum, Rev. Adv. Mater. Sci. **6**, 7 (2004).

[136] U. Brossmann, R. Würschum, U. Södervall, and H.-E. Schaefer, J. Appl. Phys. **85**, 7646 (1999).

[137] P.-L. Chen and I.-W. Chen, J. Am. Ceram. Soc. **77**, 2289 (1994).

[138] H. Matzke, J. Phys. Colloq. **34**, C9 (1973).

[139] R. J. Tesch, C. D. Wirkus, and M. F. Berard, J. Am. Ceram. Soc. **65**, 511 (1982).

[140] S. Swaroop, M. Kilo, C. Argirusis, G. Borchardt, and A. H. Chokshi, Acta Mater. **53**, 4975 (2005).

[141] M. Schie, S. Menzel, J. Robertson, R. Waser, and R. A. De Souza, Phys. Rev. Mater. **2**, 035002 (2018).

[142] B. Traoré, P. Blaise, and B. Sklénard, J. Phys. Chem. C **120**, 25023 (2016).

[143] V. A. Gritsenko, D. R. Islamov, T. V. Perevalov, V. S. Aliev, A. P. Yelisseyev, E. E. Lomonova, V. A. Pustovarov, and A. Chin, J. Phys. Chem. C **120**, 19980 (2016).

[144] X. Zhao and D. Vanderbilt, Phys. Rev. B **65**, 233106 (2002).

[145] P. Kofstad and D. J. Ruzicka, J. Electrochem. Soc. **110**, 181 (1963).

[146] N. M. Tallan, W. C. Tripp, and R. W. Vest, J. Am. Ceram. Soc. **50**, 279 (1967).

[147] A. Guillot and A. M. Anthony, J. Solid State Chem. **15**, 89 (1975).

[148] V. V. Kharton, A. A. Yaremchenko, E. N. Naumovich, and F. M. B. Marques, J. Solid State Electrochem. **4**, 243 (2000).

[149] C. Ko, M. Shandalov, P. C. McIntyre, and S. Ramanathan, Appl. Phys. Lett. **97**, 082102 (2010).

[150] A. S. Foster, F. L. Gejo, A. L. Shluger, and R. M. Nieminen, Phys. Rev. B **65** (2002).

[151] P. Broqvist and A. Pasquarello, Appl. Phys. Lett. **89**, 262904 (2006).

[152] J. L. Gavartin, D. M. Ramo, A. L. Shluger, G. Bersuker, and B. H. Lee, Appl. Phys. Lett. **89**, 082908 (2006).

[153] K. Xiong, J. Robertson, M. C. Gibson, and S. J. Clark, Appl. Phys. Lett. **87**, 183505 (2005).

[154] W. L. Scopel, A. Da Silva, JR, W. Orellana, and A. Fazzio, Appl. Phys. Lett. **84**, 1492 (2004).

[155] A. Kerber, E. Cartier, L. Pantisano, R. Degraeve, T. Kauerauf, Y. Kim, A. Hou, G. Groeseneken, H. E. Maes, and U. Schwalke, IEEE Electron Device Lett. **24**, 87 (2003).

[156] H. Takeuchi, D. Ha, and T.-J. King, J. Vac. Sci. Technol. A **22**, 1337 (2004).

[157] G. Ribes, J. Mitard, M. Denais, S. Bruyere, F. Monsieur, C. Parthasarathy, E. Vincent, and G. Ghibaudo, IEEE Trans. Device Mater. Relib. **5**, 5 (2005).

[158] J. Mitard, C. Leroux, G. Reimbold, X. Garros, F. Martin, and G. Ghibaudo, in *Defects in High-k Gate Dielectric Stacks*, NATO Science Series II, Vol. 220, edited by E. Gusev (Springer, Dordrecht, 2006) pp. 73–84.

[159] K. Sasaki and J. Maier, Solid State Ionics **134**, 303 (2000).

[160] D. Duncan, B. Magyari-Köpe, and Y. Nishi, IEEE Electron Device Lett. **37**, 400 (2016).

[161] M. Nakayama, H. Ohshima, M. Nogami, and M. Martin, Phys. Chem. Chem. Phys. **14**, 6079 (2012).